# Smart Innovation, Systems and Technologies

Volume 257

The Smart Innovation, Systems and Technologies book series encompasses the topics of knowledge, intelligence, innovation and sustainability. The aim of the series is to make available a platform for the publication of books on all aspects of single and multi-disciplinary research on these themes in order to make the latest results available in a readily-accessible form. Volumes on interdisciplinary research combining two or more of these areas is particularly sought.

The series covers systems and paradigms that employ knowledge and intelligence in a broad sense. Its scope is systems having embedded knowledge and intelligence, which may be applied to the solution of world problems in industry, the environment and the community. It also focusses on the knowledge-transfer methodologies and innovation strategies employed to make this happen effectively. The combination of intelligent systems tools and a broad range of applications introduces a need for a synergy of disciplines from science, technology, business and the humanities. The series will include conference proceedings, edited collections, monographs, handbooks, reference books, and other relevant types of book in areas of science and technology where smart systems and technologies can offer innovative solutions.

High quality content is an essential feature for all book proposals accepted for the series. It is expected that editors of all accepted volumes will ensure that contributions are subjected to an appropriate level of reviewing process and adhere to KES quality principles.

Indexed by SCOPUS, EI Compendex, INSPEC, WTI Frankfurt eG, zbMATH, Japanese Science and Technology Agency (JST), SCImago, DBLP.

All books published in the series are submitted for consideration in Web of Science.

More information about this series at https://link.springer.com/bookseries/8767

Lakhmi C. Jain · Roumen Kountchev · Bin Hu ·
Roumiana Kountcheva
Editors

# Smart Communications, Intelligent Algorithms and Interactive Methods

## Proceedings of 4th International Conference on Wireless Communications and Applications (ICWCA 2020)

Springer

*Editors*
Lakhmi C. Jain
KES International
Selby, UK

Roumen Kountchev
Faculty of Telecommunications
Technical University of Sofia
Sofia, Bulgaria

Bin Hu
Changsha Normal University
Changsha, Hunan, China

Roumiana Kountcheva
TK Engineering
Sofia, Bulgaria

ISSN 2190-3018 ISSN 2190-3026 (electronic)
Smart Innovation, Systems and Technologies
ISBN 978-981-16-5163-2 ISBN 978-981-16-5164-9 (eBook)
https://doi.org/10.1007/978-981-16-5164-9



This Springer imprint is published by the registered company Springer Nature Singapore Pte Ltd.
The registered company address is: 152 Beach Road, #21-01/04 Gateway East, Singapore 189721, Singapore

# Preface

This book is the first volume of the papers presented at the 4th International Conference on Wireless Communications and Applications (ICWCA2020) which was carried out on December 18–20, 2020, at Hainan University, Sanya, China. The papers cover the contemporary areas of smart communications, intelligent algorithms and interactive methods and related applications and present the latest achievements of the authors.

The approach of the authors is based in general on the analysis and processing of multiple signals and events data obtained from various networks and wireless sensors and performed by using a great diversity of up-to-date tools: big data analysis, neural networks, contemporary architectures, algorithms, deep learning, and augmented reality.

Special attention attracts the presented applications based on novel approaches in the areas, which give a reliable and convincing impact in the contemporary quality of life, such as the application of artificial intelligence in wireless sensor network coverage; design of cooperative driving systems; multi-user communication algorithm; intelligent rail-guided vehicle and computer numerical control; application of big data in management of distribution network operation; intelligent vehicle verification system; remote intelligent guidance system based on augmented reality; intelligent information collection, remote teaching, and many others.

The aim of the book is to present the latest achievements of the authors to a wide range of readers, namely IT specialists, engineers, physicians, Ph.D. students, and other specialists.

Selby, UK
Sofia, Bulgaria
Changsha, China
Sofia, Bulgaria
June 2021

Lakhmi C. Jain
Roumen Kountchev
Bin Hu
Roumiana Kountcheva

**Acknowledgments** The book editors express their special thanks to chapter reviewers for their efforts and goodwill to help for the successful preparation of the book. Special thanks for Prof. Dr. Lakhmi C. Jain (Honorary Chair), Prof. Dr. Srikanta Patnaik (Program Chair), Prof. Chong Shen and Prof. Kun Zhang (General Chairs), Prof. Dr. Madjid Tavana (International Advisory Chair), and Silai Zhou (Organizing Chair) of ICWCA2020.

The editors express their warmest thanks to the excellent Springer team which made this book possible.

# Contents

# About the Editors

**Prof. Dr. Lakhmi C. Jain,** PhD, Dr H.C., ME, BE (Hons), Fellow (Engineers Australia), is with the Liverpool Hope University and the University of Arad. He was formerly with the University of Technology Sydney, the University of Canberra and Bournemouth University.

Professor Jain founded the KES International for providing a professional community the opportunities for publications, knowledge exchange, cooperation and teaming. Involving around 5,000 researchers drawn from universities and companies world-wide, KES facilitates international cooperation and generate synergy in teaching and research. KES regularly provides networking opportunities for professional community through one of the largest conferences of its kind in the area of KES. http://www.kesinternational.org/organisation.php

His interests focus on the artificial intelligence paradigms and their applications in complex systems, security, e-education, e-healthcare, unmanned air vehicles and intelligent agents.

**Prof. Dr. Roumen Kountchev,** D.Sc. is with the Faculty of Telecommunications, Dept. of Radio Communications and Video Technologies—Technical University of Sofia, Bulgaria. He has 434 papers published in magazines and conference proceedings, 18 books, and 22 patents. A member of Euro Mediterranean Academy of Arts and Sciences; President of Bulgarian Association for Pattern Recognition; Editorial board member of IJBST Journal Group; Editorial board member of: Intern. J. of Reasoning-based Intelligent Systems; Intern. J. Broad Research in Artificial Intelligence and Neuroscience; Editor of books for Springer SIST series.

**Bin Hu** is an engineer and a counselor at Changsha Normal University, ACM Member, IEEE Member, GCDF, he is the guest editor of EI-indexed journals or book series: International Journal of Reasoning-based Intelligent Systems (IJRIS), International Journal of Information Systems and Supply Chain Management (IJISSCM), International Journal of Information Systems in the Service Sector (IJISSS) and Smart Innovation, Systems and Technologies (SIST), the technical editor of International Journal of Computer and Communication Technology (IJCCT). He published many articles in high standard journals like Knowledge-based System (KS), Journal of Intelligent and Fuzzy Control (JIFS), Enterprise Information System (EIS), Financial Innovation (FI) and etc.

**Dr. Roumiana Kountcheva,** S.R. is Vice President of TK Engineering. She got her PhD at the Technical University of Sofia and became Senior Researcher at TIE, in 1993. She has 183 publications and 5 patents, and presented 16 plenary speeches at international conferences and workshops. She is a member of IRIEM, IDSAI, IJBST Journal Group, and Bulgarian Association for Pattern Recognition. She is a reviewer of WSEAS conferences and journals, and of books for Springer SIST series.

# Chapter 1
# Algorithm and Application of Artificial Intelligence in Wireless Sensor Network Coverage

**Ruiyang Gao, Tingran Chen, Xin Zhang, and Ziliang Lv**

**Abstract** Artificial intelligence algorithm with its powerful processing ability to solve existing problems has attracted the attention of many experts and scholars. It has been gradually combined with the infinite sensor network, and a lot of research has been done on the energy control positioning system, router selection, and so on. Due to the slow convergence of particle swarm algorithm is easy to be the best to solve these problems, this paper presents a particle swarm algorithm based on artificial potential field (virtual force particle swarm optimization algorithm, VFPSO), in the artificial potential field algorithm, particle swarm optimization algorithm of iterative process introduced virtual force between particles, and the initial solution of optimization problem is more evenly spread out, the front-end optimization iteration of repulsive force is stronger. In the later stage, the anchor nodes are enhanced to improve the convergence efficiency of the algorithm, and the best solution with higher quality is found.

## 1.1 Wireless Sensor Network Application and Coverage

With the rapid development of micro-electromechanical system, wireless communication, information network, and integrated circuit, the development of wireless sensor network technology formed on this basis also advances by leaps and bounds. Most of the sensor nodes are installed by air drop and other selection methods, which are used for monitoring in a large range of monitoring areas or hostile parties or dangerous areas [1, 2]. Wireless sensor network (WSN) can be simulated as consisting of three modules: data information acquisition network, data information dispersion network, and control center, among which the main element is sensor node. The sensor node is generally composed of data information collection, data information disposal, data information transmission, and power supply. Infinite sensor nodes can be independent of the network setup, used for data collection at the same time by the routing protocol (such as dance by single or multiple relay routing) will be

R. Gao (✉) · T. Chen · X. Zhang · Z. Lv
Beihang University, Beijing, China
e-mail: baculuu@buaa.edu.cn



L. C. Jain et al. (eds.), *Smart Communications, Intelligent Algorithms and Interactive Methods*, Smart Innovation, Systems and Technologies 257,
https://doi.org/10.1007/978-981-16-5164-9_1

collected data transmission of information to the information data acquisition nodes (sink), data acquisition node and data disposal by the wide area network (such as the Internet or satellite network) to achieve communication with the remote center, to collect the information for treatment [3].

## 1.2 Overview of Particle Swarm Optimization Based on Artificial Potential Field

Particle swarm optimization algorithm is a biological population activity behavior originated from birds. In the process of bird seeking for food, the whole bird group maintains the unity, and individuals also retain the common mechanism of information, so that the bird group will not be excessively gathered and scattered. Since the proposal of algorithm was issued, researchers have focused on optimizing the function and strategy of algorithm, and at the same time, it is widely used in many industries such as update problem, automatic control and information mining, etc. [4].

For a long time, many scholars have carried out a wide range of research and analysis on it and gained a lot of progress. The basic idea of artificial potential field method is to simulate the influence of fictitious force field on the robot during its movement, and the movement direction of the robot is the combination of the influence of the attraction of the target object and the exclusion force of the blocking object. The mobile robot is simplified as a particle in two-dimensional space, and the movement direction $X$ ($X = (X, y)$) at any orientation in the operating space is the total field strength direction of the repulsive field of the object and the attraction field of the target.

Because of its practical simplicity and high energy effect, artificial potential field method has been widely used in the robot path planning industry. The basic thinking of artificial potential field is that the robot in the path environment can judge its running track by perceiving the attractive potential field of the target point and the crowding force field of the obstructions.

In the environment, $Q$ represents the area where the object is located, QR represents the target area, $U(Q)$ represents the simulated potential field, so the virtual gravitational potential field is represented as $U_{\text{att}}(Q)$, and the virtual repulsive potential field is represented as $U_{\text{rep}}(Q)$.

The virtual potential field and virtual force received by the robot in a certain region are shown as follows (the virtual force can be obtained by differentiating the virtual potential field)

$$U(q) = U_{\text{att}}(q) + U_{\text{rep}}(q)$$
$$\vec{F}(q) = -\nabla U(q) = -\nabla U_{\text{att}}(q) - \nabla U_{\text{rep}}(q)$$

Its $\Delta U(q)$ represents the gradient of $U$ at $Q$, which is a vector, where the direction represented is the direction with the greatest rate of change of potential field at $Q$. For position $q$ of $x$ and $y$ in two dimensions,

$$\nabla U(q) = \begin{cases} \frac{\partial U}{\partial x} \\ \frac{\partial U}{\partial y} \end{cases}$$

The minimum distance from the robot to the obstacle is $P(q)$. $\xi/\eta$ is the proportional coefficient. The fictional power obtained is:

$$\overrightarrow{F_{\text{att}}}(q) = -\xi(q - q_g)$$
$$\overrightarrow{F_{\text{rep}}}(q) = \frac{\eta}{p^2(q)}\left(\frac{1}{p(q)} - \frac{1}{p_0}\right)\nabla p(q)$$

Finally, the sum can be calculated by the crowding force and attraction force accumulated on the robot, and the movement direction of the robot can be determined according to the sum force.

Virtual force algorithm is an artificial potential field method. All nodes are treated as charged particles and there are interaction forces between nodes. When the distance between nodes is less than a certain threshold, repulsive forces are generated between nodes. When the distance between nodes is greater than the threshold, the attraction between nodes is generated. The virtual force accepted by a node is the resultant force generated by all nodes, and the resultant force is used as the direction to guide the node to move. Within the area coverage, nodes can be dispersed from a dense area more quickly to achieve a uniform distribution effect, as shown in Fig. 1.1. If $S_1$ is a mobile node, $S_2$, $S_3$, and $S_5$ are all within the communication range of $S_1$; there is a virtual force between these nodes and $S_1$, and $S_4$ is not within the communication range, so there is no force on $S_1$. If the neighbor nodes of $S_1$, $S_2$, and $S_3$ on $S_1$

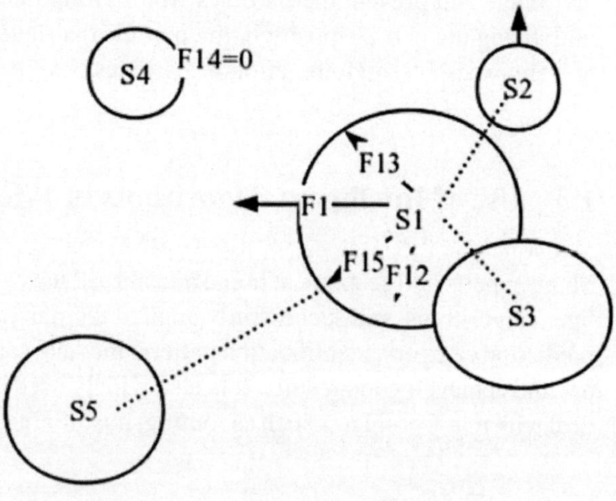


**Fig. 1.1** Schematic diagram of virtual force action

repulsion $F_{12}$ and $F_{13}$, $S_5$ and $S_1$ are far apart but produce less attractive $F_{15}$, $S_4$ does not produce force, then $S_1$ by force of $F_1$, $F_{12}$, $F_{13}$, $F_{15}$, the sum of the power of the mobile node $S_1$ along the resultant force direction from the current position to move to a new location.

Simulation $S = \{S_1, S_2, S_3, \ldots, S_N\}$ is node set, so node $i$ and $j$ are $S_i$ and $S_j$, the distance $d_{ij}$ between them is as follows:

$$d_{ij} = d(s_i, s_j) = \sqrt{(x_i - x_j)^2 + (y_i - y_j)^2}$$

Therefore, the force of node $S_j$ on $S_i$ is formulated as follows:

$$\overrightarrow{F_{ij}} = \begin{cases} 0, & \text{if, } d_{ij} > R_c \\ (w_a(d_{ij} - D_{\text{the}}), \alpha_{ij}), & \text{if, } R_c \geq d_{ij} > D_{\text{the}} \\ 0, & \text{if, } d_{ij} = D_{\text{the}} \\ (w_r(D_{si} - d_{ij}), \alpha_{ij} + \pi, & \text{if, } d_{ij} < D_{\text{the}} \end{cases}$$

$d_{ij}$ — $S_j$ on $S$ Euclidean distance.
$a_{ij}$ — $S_j$ on $S$ direction angle of line segment.
$\omega_a/\omega_r$ — a virtual force gravitational force repulsive force parameter.
$D_{\text{the}}$ — Set distance threshold.
$R_c$ — the communication radius of the node.

The resultant force exerted on the node is as follows:

$$\overrightarrow{F_i} = \sum_{j \in S} \overrightarrow{F_{ij}}$$

In the process of particle swarm optimization, the gravitational force and repulsive force are introduced into the robot in the artificial potential field method, and the virtual force generated between the particles is used to enlarge the particle optimization range and prevent the particles from falling into the local optimal value. After initializing the particle position, the particle distribution is more uniform in order to prevent excessive agglomeration of particles.

## 1.3 Use of Intelligent Algorithms in WSN Coverage

Many experts and scholars at home and abroad have realized the advantages of intelligent algorithms, and successfully applied them in data disposal, automatic control, prediction category, identification pattern, medical category, military category, and so on, and achieved good results. It is also widely used in the infinite sensor network to deal with many problems such as routing, positioning, coverage selection, and energy control, among which the application of intelligent algorithm in WSN coverage is mainly studied [5].

Particle swarm optimization is a traditional intelligent optimization algorithm. Since its introduction, it has attracted the attention of many researchers. At the same time, it has been applied in many researches and has grown into a corresponding separate discipline, which is applied in intelligence improvement, information discovery and signal disposal. It is widely used in many fields such as automation control, flow shop management, and network transmission.

Different from other intelligent algorithms, particle swarm optimization focuses more on comprehensive retrieval. In the process of iteration, particle swarm optimization can reach a certain category and finally obtain the optimal solution. It has a strong robustness that makes it easier to operate. In the optimization process, this algorithm is easy to be troubled by the local optimization when solving the problem with high complexity and fall into the iterative process of infinite loop, which does not hinder the retrieval of the optimal solution. The first factor that falls into the infinite loop is that the global optimal value and the historical optimal value of each particle in the iteration process will affect the update of particle position. Although the initial motion will be improved under the action of algorithm convergence, the later algorithm is difficult to break and seek for the optimal solution, so it is difficult to get rid of the trouble of local optimal. After studying and practicing particle swarm optimization algorithm and other intelligent algorithms, the limitations of the algorithm are obviously improved, the original defects of the algorithm are overcome, and the optimal solution or more perfect solution can be obtained through the optimized algorithm. Now integrating particle swarm optimization with other intelligent algorithms has become an important optimization method to improve particle swarm optimization.

## 1.4 Conclusion

This paper improves the standard particle swarm algorithm, the first artificial potential field algorithm and the convergence of particle swarm optimization, and take charge of the interaction of the introduction of particles in the optimization process of thinking, try to use the optimal particle in the iteration process for the target point and exclude force generated between particles avoid algorithm falls into local optimum, and finally to seek the optimal solution.

Particle swarm optimization promotes particle average dispersion, which can effectively avoid excessive particle convergence and make the algorithm fall into the local optimal solution. However, due to the large amount of calculated work, it is more suitable to solve the problem of small coverage and obtain a better layout. Due to its simplicity, particle swarm optimization algorithm is more suitable for solving large area coverage problems. Compared with standard particle swarm optimization algorithm, particle swarm optimization algorithm can obtain the optimal network layout more efficiently.

## References


1. Jiang, J.R., Chen, Y.C., Lin, T.Y.: Particle swarm optimization for charger deployment in wireless rechargeable sensor networks. Int. J. Parallel Emergent Distrib. Syst., 1–16 (2018)
2. Mahboubi, H., Aghdam, A.G.: Distributed deployment algorithms for coverage improvement in a network of wireless mobile sensors: relocation by virtual force. IEEE Trans. Control Netw. Syst. **4**(4), 736–748 (2017)
3. Ma, F., Zhang, L., Wang, J.B.: Improvement of particle swarm optimization algorithm and its application in optimization function. Comput. Digit. Eng. **7**, 1252–1293 (2017)
4. Frans, V.D.B.: An analysis of particle swarm optimizers. University of Pretoria (2002)
5. Jain, N.K., Nangia, U., Jain, J.: A review of particle swarm optimization. J. Inst. Eng. **3**, 1–5 (2018)

# Chapter 2
# Design of Cooperative Driving System Based on VANET

**Jing Tao Li, Xiao Jiang Li, Zhen Feng Li, and Run Qiu**

**Abstract** The rapid development of automobile industry has brought great convenience to people's travel, but the increase of vehicles also brings a series of environmental problems and traffic problems, so this paper designs a cooperative driving system based on vehicular ad hoc network (VANET). As a product of the integration of communication industry and automobile industry, VANET can provide drivers with the perception ability beyond visual range, thus giving drivers more reaction time and reducing the probability of traffic accidents. In addition, in some specific environment, the system can achieve a certain degree of automatic driving, reduce the driver's fatigue, but also reduce fuel consumption. Finally, this paper uses Simulation of Urban MObility (SUMO) to simulate the vehicle driving at the crossroads. According to the simulation results, under the action of the system, the vehicles can pass through the intersection safely and effectively.

## 2.1 Introduction

Every year, many people all over the world die in traffic accidents. Most of these accidents are caused by the driver's wrong judgment. With the birth of the vehicle auxiliary driving system, many sensors are installed on the body, which helps the driver to make decisions to a certain extent and reduces the incidence of traffic accidents. But only using the sensor, the sensing distance of the car is relatively close, so it cannot play a good role of the sensor. Sometimes, it may even cause the driver's misjudgment because of the false alarm of the sensor.

As a new technology, VANET has integrated automobile industry and communication. It can interconnect the vehicles on the road with vehicles, between vehicles and people and between vehicles and fixed nodes, to form an easy to deploy and cheap intervehicle communication network [1]. The network can effectively manage the traffic flow, make traffic more fluid, and reduce traffic congestion and the occurrence of the accident. Because of the characteristics of self-organization, no center, and

J. T. Li (✉) · X. J. Li · Z. F. Li · R. Qiu
Faculty of Information Engineering and Automation, Kunming University of Science and Technology, Kunming 650051, China

 
L. C. Jain et al. (eds.), *Smart Communications, Intelligent Algorithms and Interactive Methods*, Smart Innovation, Systems and Technologies 257,
https://doi.org/10.1007/978-981-16-5164-9_2

supporting multihop forwarding of data, the application of vehicular ad hoc network in cooperative driving system can provide drivers with the perception ability beyond visual range, which can provide more time for drivers to deal with some emergency situations [2]. At the same time, it can also provide a series of functions such as traffic accident early warning, traffic congestion query, and path planning. In addition, because the vehicle is connected to the Internet, it can also provide more entertainment for the passengers on the vehicle.

## 2.2 Overall System Model

The system is composed of road test system, vehicle system, and human–computer interaction interface. The overall structure of the cooperative driving system is shown in Fig. 2.1.

The system is composed of road test system, on-board system, and human–computer interface. When the system is in operation, on the one hand, the road test system collects traffic lights within a certain range and at the same time counts the number of current road vehicles and numbers these vehicles and calculates the average speed and flow density of vehicles in the current area. On the other hand, it combines the height of the system precision map, real-time monitoring of road conditions, and then the combination of the two information sent to the road test control unit, and the road test control unit makes decisions according to the information obtained and judges the crowded section and accident emergency section

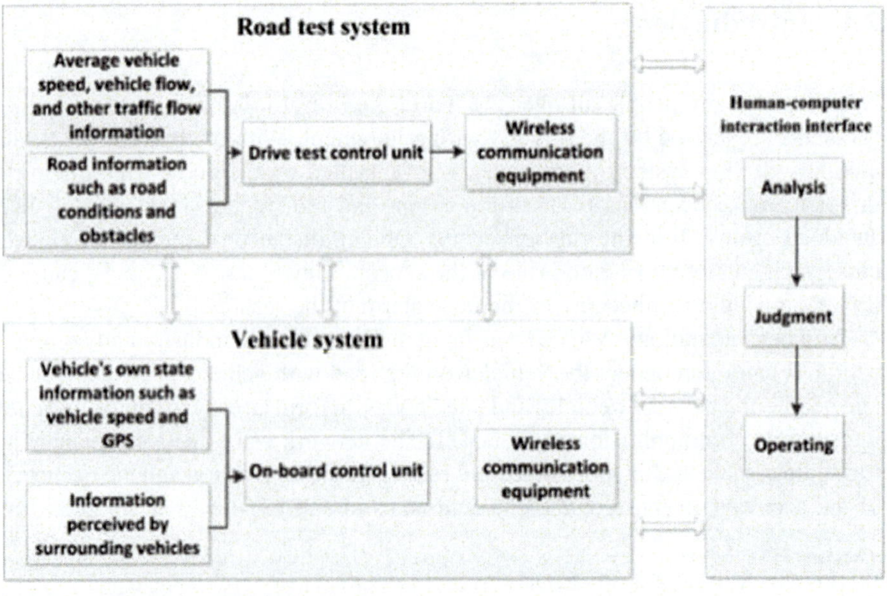


**Fig. 2.1** Overall structure of the cooperative driving system

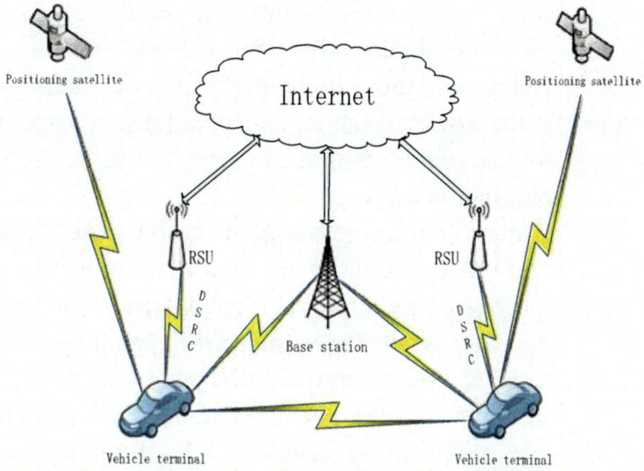


**Fig. 2.2** Structure diagram of the collaborative driving system

in the area. Finally, the decision-making information is broadcast to the vehicles in the area through dedicated short range communication (DSRC) [3]. The on-board system records the vehicle's own speed, acceleration, angle, and other motion state information through the speed sensor and inertial sensor. At the same time, it displays the geographic location of the current vehicle, such as roads and lanes, in combination with the vehicle's own high precision map and global positioning system (GPS) location information. The DSRC can also continuously share the information of vehicle perception between vehicles to improve the perception range of vehicles. Then, the information of vehicle motion state and the perception information of surrounding vehicles are sent to the on-board control unit for data analysis and decision-making, so as to judge whether the current vehicle changes lanes, accelerates, and decelerates. Finally, the decision-making information is transmitted to the human–computer interface through controller area network bus. The driver analyzes the decision-making and perception information sent by the road test system and on-board system to determine whether the current vehicle is changing lanes, accelerating, decelerating, or following. The network structure diagram of the collaborative driving system is shown in Fig. 2.2.

## 2.3 Working Principle of the Collaborative Driving System

### *2.3.1 Vehicle Lane Change*

If a vehicle wants to leave the current road, it first needs to perceive the surrounding environment and traffic conditions of the vehicle, then reduce the speed, and enter the turning lane. The steps of lane changing are as follows:

(1) When vehicles need to change lanes, vehicles broadcast lane change information and their own speed and other dynamic parameter information to

surrounding vehicles through vehicle ad hoc network. Other vehicles will judge the distance between the two vehicles after receiving the information and remind the vehicle driver to drive carefully if the distance is close.

(2) If the driver of the rear vehicle fails to respond to the warning message sent by the vehicle within the specified time, the vehicle will automatically take braking measures.

(3) If there is a traffic accident on the road, the vehicles in the ad hoc network will broadcast traffic accident news to nearby vehicles. At the same time, the location of the traffic accident will be located according to the road test unit and the place where the vehicle GPS signal last appeared, helping the rescue team to be able to reach the place where the accident occurred in time for rescue. In addition, the vehicles near the traffic accident can transmit the warning information to the farther vehicles through the DSRC multihop communication technology [4]. After receiving the warning message, the remote vehicles will try their best to find other routes to avoid driving into the traffic accident section.

### *2.3.2 Vehicle Following*

When a vehicle is driving on a highway, the speed of the vehicle tends to remain the same for a period of time, so the vehicle can often choose to follow the car, so that the fatigue of driving can be effectively resolved. The vehicle uses the high precision map and GPS information to calculate the distance between the vehicle ahead and the current vehicle. At the same time, proportion integral differential (PID) control algorithm is used to synchronize the speed of the current vehicle with the speed of the vehicle in front, so that the distance between the front and rear vehicles is always kept at a safe distance [5]. When the distance between the front and rear vehicles is less than the safe distance, the warning function can be activated. When the driver fails to realize the danger or ignores the warning information, it can automatically perform the brake function to ensure the driving safety. The following steps are as follows:

(1) The vehicle obtains external environment information through its own perception, and information exchange in the vehicle ad hoc network uses high precision maps and GPS to detect the distance between the current vehicle and the vehicle in front and uses visual sensors and image processing technology to determine the lane of the vehicle in front.

(2) Through the vehicle sensor to obtain the current vehicle speed, acceleration and other driving state information and the driver's operation status.

(3) According to the status information obtained, the car-following system will be withdrawn when the current vehicle or the current vehicle needs to drive out of the current road. At the same time, the risk assessment of collision between the front and rear vehicles is carried out. The assessment result will be sent to the early warning system and brake control system, and the driver will send out early warning information through the alarm when there is a risk of collision.

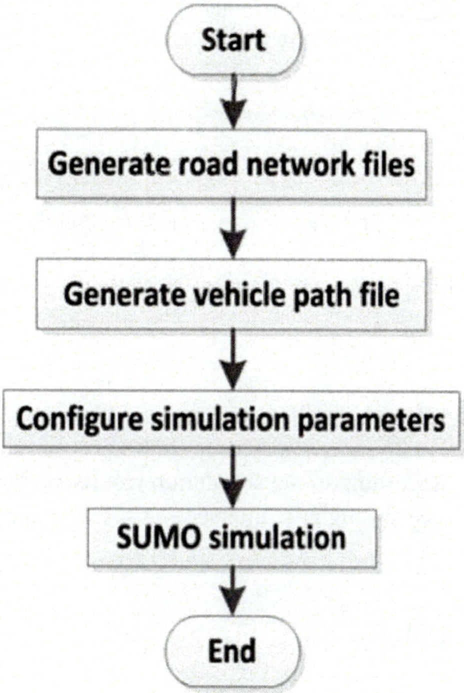


**Fig. 2.3** Simulation flowchart

## 2.4 System Simulation

In all traffic accidents, the intersection is the place where traffic accidents happen most easily, so this paper uses SUMO simulation software to generate a crossroad intersection and generate a number of vehicles on the road. The simulation flowchart is shown in Fig. 2.3.

The goal of simulation is to transmit the turning intention of each vehicle to other vehicles at the intersection through vehicular ad hoc network, so as to realize the safe and effective passage of vehicles through the intersection. The simulation results show that the cooperative driving system based on VANET can effectively improve the driving safety, and the simulation results are shown in Fig. 2.4.

## 2.5 Conclusion

In order to solve the problem of vehicle driving safety, this paper designs a cooperative driving system based on VANET. The system is based on wireless ad hoc network, which expands the driver's perception distance, and has a certain warning function for the coming traffic accidents, so as to give the driver more reaction time and avoid the occurrence of traffic accidents. At the same time, in some specific scenarios, the system can achieve a certain degree of automatic driving, so it can effectively

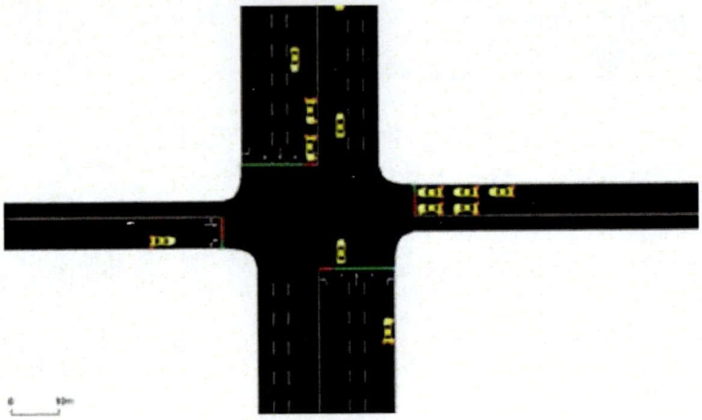

**Fig. 2.4** Simulation results of the collaborative driving system

help drivers alleviate driving fatigue and reduce the probability of traffic accidents. Finally, this paper uses SUMO to simulate the vehicle driving at the crossroads. According to the simulation results, under the action of the system, the vehicles can pass through the intersection safely and effectively.

## References

1. Dou, Z.X., Wang, X.N.: Data delivery in VANET with network gaps. J. High Speed Netw., 111–120 (2020)
2. Kumar, N., Kaur, K., Jindal, A., Rodrigues, J.J.: Providing healthcare services on-the-fly using multi-player cooperation game theory in Internet of Vehicles (IoV) environment. Digit. Commun. Netw., 191–203 (2015)
3. Raza, H., Ioannou, P.: Vehicle-following control design for automated highway systems. IEEE Trans. Control Syst. Technol., 44–57 (1996)
4. Van Dijck, T., van der Heijden, G.A.: Vision sense: an advanced lateral collision warning system. In: IEEE Proceedings, Intelligent Vehicles Symposium, pp. 290–302 (2005)
5. Jia, D.Y., Zhang, R., Lu, K.J., Wang, J.P.: Improving the uplink performance of drive-thru internet via platoon-based cooperative retransmission. IEEE Trans. Veh. Technol., 4536–4545 (2014)

# Chapter 3
# A Study of Intelligent Power Information Collection System Based on Distributed Architecture

**Bin Sun, Haiying Li, Ziping Long, Hua Zhang, Zhiliang Zhao, Huaizhen Jiang, and Lirong Pu**

**Abstract** In allusion to a series of flaws in conventional power information collection system such as weak compatibility between terminals, communication blindness of collection points, low level of acquisition system integration, non-synchronous data checking and processing, and poor security protection performance, this paper puts forward and builds an intelligent power information collection system based on a three-tier distributed architecture which has main station layer, communication layer, and acquisition layer. Its purposes are to realize the following functions: automatic collecting power information, abnormal monitoring, power quality monitoring, power consumption analysis and management, information distribution, distributed power supervision, information interaction between smart power equipment, etc. With features of technical advancement and economic rationality, this system has been proved to show a way to upgrade project of rural power grid of Yunnan Agricultural Reclamation Power Co., Ltd, Xishuangbanna Dai Autonomous Prefecture, Yunnan Province, China, and further create a basic development condition for construction of regional smart grid.

B. Sun (✉)
Yunnan State Farms Group Co. Ltd., Kunming 650051, Yunnan, China
e-mail: ynkmben@163.com

H. Li
School of Foreign Languages, Yunnan Minzu University, Kunming 650504, Yunnan, China

Z. Long · H. Zhang · Z. Zhao
Yunnan Agricultural Reclamation Power Co. Ltd., Jinghong 666100, Xishuangbanna, Yunnan, China

H. Jiang
XJ Electric Co. Ltd., Xuchang 461000, Henan, China

L. Pu
Zhuhai Pilot Technology Co. Ltd., Zhuhai 519000, Guangdong, China

 
L. C. Jain et al. (eds.), *Smart Communications, Intelligent Algorithms and Interactive Methods*, Smart Innovation, Systems and Technologies 257,
https://doi.org/10.1007/978-981-16-5164-9_3

## 3.1 Introduction

Already known are a series of problems in traditional power information collection system in Refs. [1–4]. First, the compatibility between terminals is weak. Being designed to access various types of collecting terminals, communication resolution and operation management between terminals with different interface standards are problems hard to deal with for traditional systems. Most data of the existing power information collection system are directly extracted from the marketing business system, and the inconsistency of real-time data affects also the accuracy of data collection. Human-caused data distortion in the absence of conditions for collecting and copying data and problems of the inability to store and extract data effectively in the conditions for collecting and copying data are to be solved urgently. Second, the communication of collection points is blind. With the continuous expansion of power grid construction area, the number of terminals connected to the power user electricity information acquisition system is becoming larger and larger, and the amount of collected data is larger too, so to uploading and storing all the collected data in real time are necessary. Anyway, the large number of substations, lines, and substation areas corresponding to a large number of newly added new acquisition points cannot match the original system standard. Third, the level of acquisition system integration is low. The present electric energy information acquisition system meets the needs of automatic meter reading, load control, and distribution transformer monitoring, but the data interaction with marketing business module, dispatching automation module, and distribution network automation module is not realized. So, system integration is a key problem to be solved urgently. Fourth, the data checking and processing are non-synchronous. Data recognition, judgment, and processing delay under meter code, power, voltage, current, power quality, and abnormal condition events largely reduce the availability of real-time data. Fifth, the security protection performance is poor. The system collects a large amount of data; simultaneously, the data backup, archiving, disaster recovery, and other relative strategies are particularly important. Effective strategic security management, real-time data, historical data testing, cost control function of low-voltage meter reading data, and terminal side data statistics and analysis are also included. What is clear from the above is that solving some existing problems in traditional system is imperative and creating a new system architecture is urgent. Based on discussions in Refs. [5–7], this paper aims to build an intelligent power information collection platform based on distributed architecture with features of much better customer satisfaction and power sense.

## 3.2 The System Architecture

### *3.2.1 The System Configuration*

The intelligent power consumption information acquisition platform includes master station layer, communication layer, and acquisition layer in Refs. [4, 8–14]. Its logical framework is shown in Fig. 3.1. The master station realizes data processing, analysis, storage, exchange, release, and comprehensive application platform functions; communication network, with the function of supporting multiple communication modes, realizes data transmission between terminal and master station. The acquisition layer realizes the functions of data acquisition, storage, processing, uploading, and user power consumption control.

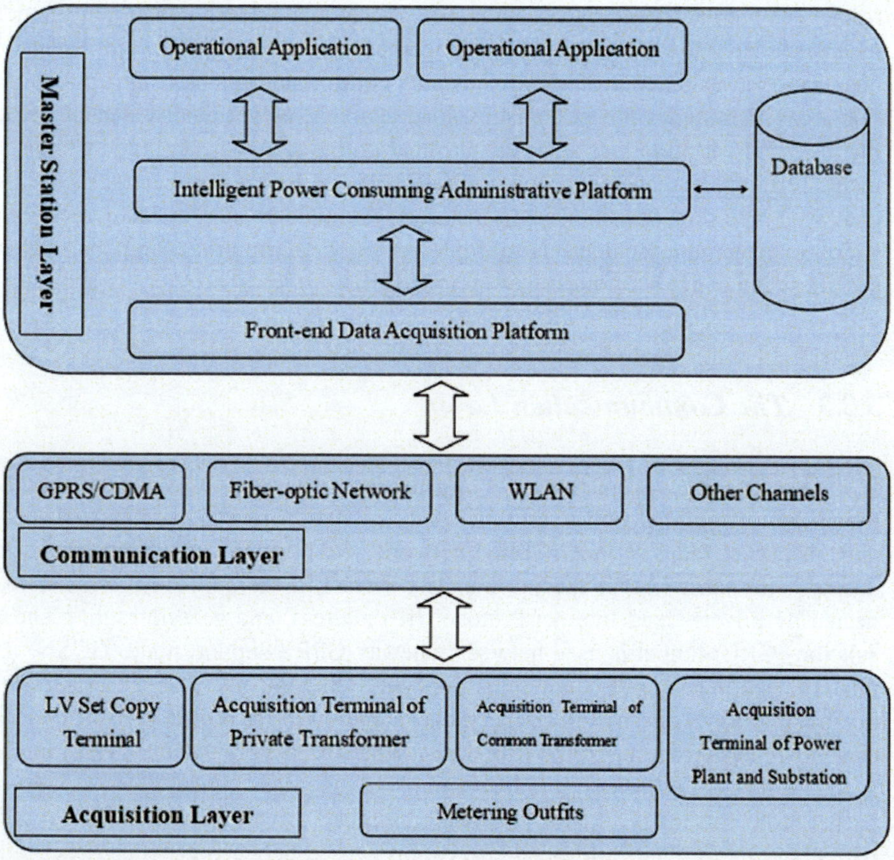


**Fig. 3.1** Logical framework of intelligent power consumption information acquisition platform

### 3.2.2 The Master Station Layer

The logic framework of the master station is constructed based on the core that contains power consumption information collection, data management and data sharing, and the focus, i.e., the application function of meeting users' demand. Therefore, the unified database platform, data acquisition platform, data management platform, and comprehensive application platform are realized.

The main station system functions include data acquisition management, business management, power management, application analysis, terminal management, and alarm management. Detailed functions are as follows. Data collection management includes implementation and current data, historical daily data, historical monthly data and event records, data rationality check, data analysis and calculation, and data storage management. The business management functions include meter reading function, meter change processing, current transformer (CT)/potential transformer (PT) replacement, bypass replacement, and manual test. Power management includes orderly power consumption management, power-fixed value control, electric quantity-fixed value control, rate-fixed value control, and remote control. Application analysis includes power statistics, load analysis, power quality data statistics, line loss and variation loss analysis, terminal online statistics, and report management. Terminal management includes basic data, terminal parameters, meter/user data, system timing, operation status management, maintenance, and fault recording. Alarm management includes alarm implementation, alarm processing, alarm event statistics, and alarm short message service (SMS).

### 3.2.3 The Communication Layer

In this paper, access point name (APN) network is selected out of the security and timeliness of regional power grid. The master station server is connected to the general packet radio service (GPRS) network of the mobile company through an access point name (APN) special line. The routers of the both sides are interconnected by private fixed Internet Protocol (IP) address, and generic routing encapsulation (GRE) tunnel is used between gateway GPRS support node (GGSN) and interconnection router of communication operation company. The subscriber identity module (SIM) card used for general packet radio service (GPRS) private network can only open the service bound with the dedicated access point name (APN) and not available for other services. After obtaining access point name (APN), all general packet radio service (GPRS) modules are assigned private fixed Internet Protocol (IP), respectively. The end-to-end encryption is used between general packet radio service (GPRS) network table and application front-end computer to prevent information leakage in the process of transmission. Both sides isolate filter Internet Protocol (IP) and port on firewall. The technical project designed in this paper is suitable for the agricultural reclamation power grid and other local power grids which are of high-security requirements, many data points, and strict timeliness requirements.

### 3.2.4 *Intelligent Terminal*

Power consumption information acquisition terminals are responsible for the power information collection, data management, two-way data transmission, and forwarding or executing control commands. The terminals have three components: distribution transformer monitoring terminal, low-tension concentrate meter reading terminal, and power consumption management terminal.

#### 3.2.4.1 The Capabilities of Distribution Transformer Monitoring Terminal

The terminal includes sub-functions like communication, data acquisition, time record reporting, and the like. The communication function includes high-efficiency data compression encryption algorithm, short message service (SMS) communication, general packet radio service (GPRS)/code division multiple access (CDMA)/public switched telephone network (PSTN)/Ethernet communication; data acquisition function includes multi-function meter data acquisition, alternating current (AC) analog acquisition, direct current (DC) analog acquisition, pulse acquisition, and switch state acquisition; event recording and reporting function includes transformer operation condition monitoring, energy meter operation status monitoring, abnormal power consumption monitoring events, and terminal self-monitoring events.

#### 3.2.4.2 The Capabilities of Low-tension Concentrate Meter Reading Terminal

The low-voltage centralized reading terminal has functions like communication, expansion, distribution transformer monitoring, and the like. The communication function *includes* high-efficiency data compression encryption algorithm, supporting communication ways of short message service (SMS)/general packet radio service (GPRS)/code division multiple access (CDMA)/public switched telephone network (PSTN)/Ethernet. The expansion function includes carrier meter reading, automatic relay, automatic meter finding, and line loss calculation as well.

## 3.3 Implementation of Intelligent Power Consumption Information Acquisition System

### *3.3.1 The Physical Architecture*

In this paper, the physical architecture based on Browser/Serve can reduce the cost of construction and maintenance and simplify the management process, too. The clients are designed web browsers and mobile apps, with a friendly application interface and simple operation, being suitable for large scale and *complex* application requirements. The component technology enables the system to have good scalability and meet the needs of building large-scale system. By adopting the cross platform design, the middle layer component provides a wide range of heterogeneous database access and replication capabilities. Therefore, automatically calculating time of use price and step price and connecting marketing system with other external systems seamlessly are realized. At the same time, with application of adopts the security design of hierarchical authority management mode and digital certificate authentication mechanism, a physical structure of intelligent electricity consumption information acquisition system is built based on Browser/Serve.

### *3.3.2 Internal and External Contact*

Normally, power grid enterprises give priority to the data acquisition system based on intrinsic safety since it meets the present needs of the current rural reclamation power grid architecture and the development requirements within a certain period in the future. The *internal* and external contact information of the system must ensure the safe transmission of information and risk-free storage of data.

## 3.4 Conclusions

The system proposed in the paper has three layers: master station layer, communication layer, and acquisition layer. It solves the problems of poor compatibility of storage terminal, blind spot of communication coverage, low integration of acquisition system, asynchronous data verification and processing, poor system security protection performance and so on. It has been proved that the system is advanced in technology and economic in Xishuangbanna, Yunnan. It can also serve as a reference for the design and construction of intelligent power information collection system under the framework of smart grid.

**Acknowledgements** This work was supported by the Golden Phoenix Talents Project in Yunnan State Farms Group Co., Ltd.

## References


1. Sun, Q.Y., Yang, L.X., Zhang, H.G.: Smart energy-applications and prospects of artificial intelligence technology in power system. Control Decis. **33**(5), 938–948 (2019)
2. Dileep, G.: A survey on smart grid technologies and applications. Renew. Energy **146**, 2589–2625 (2020)
3. Mohandes, B., Hammadi, R.A., Sanusi, W., Mezher, T., Khatib, S.E.: Advancing cyber–physical sustainability through integrated analysis of smart power systems: a case study on electric vehicles. Int. J. Crit. Infrastruct. Prot. **23**, 33–48 (2018)
4. Ansari, J., Gholami, A., Kazemi, A.: Multi-agent systems for reactive power control in smart grids. Int. J. Electr. Power Energy Syst. **83**, 411–425 (2016)
5. Sun, B., Li, H.Y., Guo, H., Li, T.: A study of smart system of power utilization safety management based on a cloud platform. In: Proceedings of the IACAE, pp. 1446–1450 (2019)
6. Sun, B., Sui, W.L., Li, H.Y.: Applied research of supervision and control system in 110 kV smart substation based on three layers of three networks. In: Proceedings of the ICIA, pp. 896–900 (2015)
7. Chen, J.C., Li, W.T., Wen, C.K., Teng, J.H., Ting, T.: Efficient identification method for power line outages in smart power grid. IEEE Trans. Power Syst. **29**(4), 1788–1800 (2014)
8. Gong, J., Zhou, S., Niu, Z.S.: Optimal power allocation for energy harvesting and power grid coexisting wireless communication system. IEEE Trans. Commun. **61**(7) (2013)
9. Didier, G., Lévêque, J., Rezzoug, A.: A novel approach to determine the optimal location of SFCL in electric power grid to improve power system stability. IEEE Trans. Power Syst. **28**(2), 978–984 (2013)
10. Mohamed, O.A., Lutz, L.: Power line communications for low-voltage power grid tomography. IEEE Trans. Commun. **12**, 5163–5175 (2013)
11. Menke, J.H., Bornhorst, N., Braun, M.: Distribution system monitoring for smart power grids with distributed generation using artificial neural networks. Int. J. Electr. Power Energy Syst. **113**, 472–480 (2019)
12. Guo, Y., Huo, R., Liu, X., Xie, Z.: Lumped parameter cable model for low voltage power line communication. Autom. Electric Power Syst. **43**(2), 156–161 (2019)
13. Alam, Z., Khursheed, A., Chaudhary, R.K.: Modeling simulation and performance evaluation of low voltage power line communication channel. Int. J. Adv. Technol. Eng. Explor. **5**(46), 308–317 (2018)
14. Saleem, U., Jangsher, S., Qureshi, H.K., Hassan, S.A.: Joint subcarrier and power allocation in energy harvesting-aided D2D communication. IEEE Trans. Ind. Inf., 99 (2018)

# Chapter 4
# A Multiuser Communication Algorithm for UAV Bee Combat

**Li Shi-ming, Yang Yuan, Wu Yi-jie, and Kang Chao**

**Abstract** The combat of unmanned aerial vehicle (UAV) bee colony is the primary trend of information warfare in the future. Under the high-dynamic environment of air-to-air combat, severe Doppler shifts generated by relative movement of UVA reduce the communication reliability between vehicles. Therefore, a new division-time multiuser chirp modulation (DT-MCM) algorithm based on chirp signal is proposed in this paper aiming to the characteristic of existing communication technology sensitive to Doppler shifts. Utilizing the insensitivity advantage of chirp signal to Doppler shift, two chirp signals with different chirp rate can constitute one chirp signal. And, our proposed method differentiates users using the chirp rate of chirp signals to implement the UAV bee colony communication. Depending on the electromagnetic environment changes, the efficient and reliable information transmission can be realized by dynamically adjusting parameters such as time and bandwidth. Applied to UAV bee colony communication and according to MATLAB simulation results, the proposed algorithm proved the effectiveness to reduce the adverse impact to UAV internetwork communication caused by Doppler shifts.

## 4.1 Introduction

Unmanned aerial vehicle (UAV) is a kind of driverless aircraft using radio remote control device and its own program-controlled apparatus [1]. UAVs are playing an important role in recent year's wars due to their small size, lightweight, flexible response, high-risk work performability, and effective reduction of combat personnel casualties [2].

As a subversive force to change the rules of future war, UAV bee colony has become the focus of research of domestic and foreign scholars [3, 4]. However, one of the key issues of UAV bee combat is the reliability between information transmission. When UAV bee colony flying, the nodes of the flight formation are close to and interact with each other as a whole. If an uncontrollable failure occurs

L. Shi-ming · Y. Yuan (✉) · W. Yi-jie · K. Chao
Shanghai Aerospace Electronic Technology Institute, Shanghai 201100, China
e-mail: yangy940402@126.com

 
L. C. Jain et al. (eds.), *Smart Communications, Intelligent Algorithms and Interactive Methods*, Smart Innovation, Systems and Technologies 257,
https://doi.org/10.1007/978-981-16-5164-9_4

at a bee colony node, it will inevitably affect other nodes, even collide. In air-to-air combat scenario, the relatively high-speed movement between vehicles will result in a large Doppler frequency shift, which will affect the colony communication [5]. Therefore, it is of great important to achieve the robust control of formation flight.

In order to better improve the reliability of information transmission between UAV networks in high-dynamic scene, this paper selects chirp signal for communication, which is suitable for transmission in selective fading channels. Chirp signal has preferable robust to Doppler frequency and multipath ability [6] due to its good time–frequency characteristics. This letter proposes a novel multi-access communication algorithm based on chirp signal between UAV networks.

## 4.2 Chirp Signal Against Doppler Frequency Shift

### *4.2.1 Chirp Signal Impulse Compression Characteristic*

One of the significant features of chirp signal is good impulse compression [7]. When it performs matched filtering with its matched signal, the output signal can be written as:

$$S(t) = \sqrt{BT}\frac{\sin\left[\pi Bt\left(1 - \frac{|t|}{T}\right)\right]}{\pi Bt}\cos(2\pi f_0 t) \tag{4.1}$$

where $S(t)$ represents the output signal, $B$ represents the signal bandwidth, and $T$ is symbol duration.

As shown in above formula, when chirp signal matched filtering, the output signal has similar performance to the sin*c* function which has a sharp autocorrelation peak. The time-domain waveform after matching modulation is shown in Fig. 4.1.

### *4.2.2 Analysis of Chirp Signal Anti-Doppler Performance*

This section analyzes the influence of the Doppler frequency shift on the impulse compression ability of the chirp signal.

Assuming that, there is only Doppler shift in the channel, and the signal arriving at the receiving end after transmission through the wireless channel is expressed as:

$$S(t) = \cos(2\pi(f_0 + f_d)t + \pi\mu t^2) \tag{4.2}$$

Fig. 4.1 Chirp matched filtering impulse compression waveform

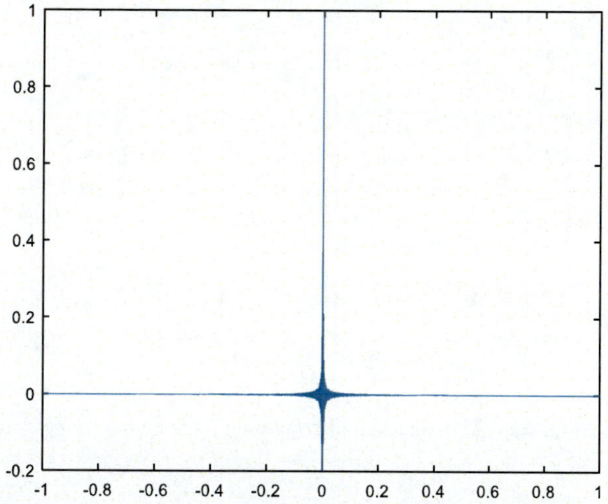


where $f_d$ represents Doppler shift and $\mu$ is chirp signal frequency modulation index.

After matched filtering received signal, the matched output signal is obtained as:

$$\begin{aligned} r(t) &= \cos\left(2\pi(f_0 + f_d)t + \pi\mu t^2\right) * \sqrt{2\mu}\cos\left(2\pi f_0 t - \pi\mu t^2\right) \\ &= \sqrt{\mu}\frac{\sin(\pi(\mu t + f_d)(T - |t|))}{\pi(\mu t + f_d)}\cos(2\pi(f_0 + f_d)t) \end{aligned} \tag{4.3}$$

It can be seen from formula (4.3) that when there is a Doppler shift in the channel and when the chirp signal matched filters with its matched signal, the output signal still has a similar performance to sinc function. However, the amplitude and peak position of the output waveform change, and the amplitude value decreases to $1 - |\frac{f_d}{\mu}|$ times. Since the chirp signal has a good time–frequency coupling characteristic, it can convert the frequency shift caused by the existence of Doppler into a time shift. The size of the time shift is $\frac{f_d}{\mu}$. Figure 4.2 is the overall diagram and the partial enlarged diagram of the output waveform of the chirp signal matched filter under different Doppler frequency shifts.

## 4.3 Division-Time Multiuser Chirp Modulation Algorithm

From the above analysis, it can be observed that chirp signal has good anti-Doppler frequency shift performance. Therefore, based on the chirp signal, this paper presents a new type of communication algorithm suitable for UAV bee network, which can not only meet the reliable information transmission in high-dynamic air-to-air combat

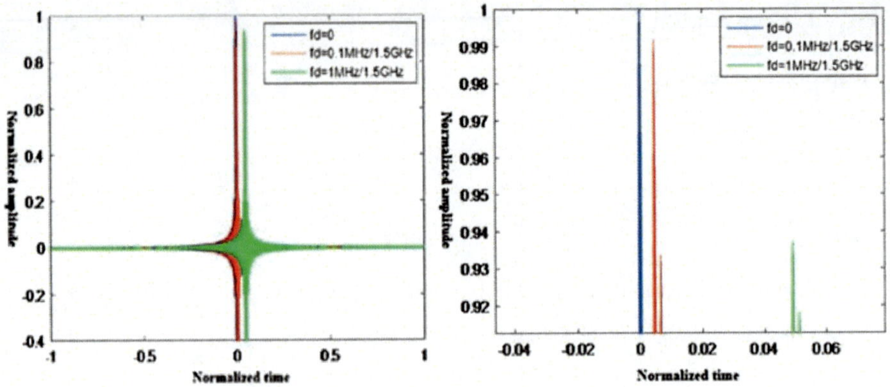


**Fig. 4.2** Chirp matched filtering impulse compression waveform under different Doppler shifts

scenarios but also save the system frequency band resources. This algorithm allocates user's modulation frequency according to the time, so it is called division-time multiuser chirp modulation (DT-MCM). The time–frequency diagram of this algorithm is shown in Fig. 4.3.

The algorithm uses chirp-BOK modulation. When the frequency modulation polarity is positive, namely $\mu > 0$, data 1 is sent, and when the frequency polarity is negative, namely $\mu < 0$, data 0 is sent [8]. Each user in the cellular network is assigned a pair of unique chirp signals composed of two frequency modulations, which are divided into two equal parts in a unified frequency band. The overall mathematical expression of the DT-MCM algorithm is as follows:

$$s_k(t) = \begin{cases} s_k^1(t) = \sqrt{\frac{2E}{T}} \cos(2\pi f_c t + \pi \mu_k^1 t^2), & 0 \le t \le t_k \\ s_k^2(t) = \sqrt{\frac{2E}{T}} \cos\left[2\pi (f_c + B/2)(t - t_k) + \pi \mu_k^2 (t - t_k)^2\right] & t_k \le t \le T \end{cases} \tag{4.4}$$

**Fig. 4.3** DT-MCM time–frequency diagram

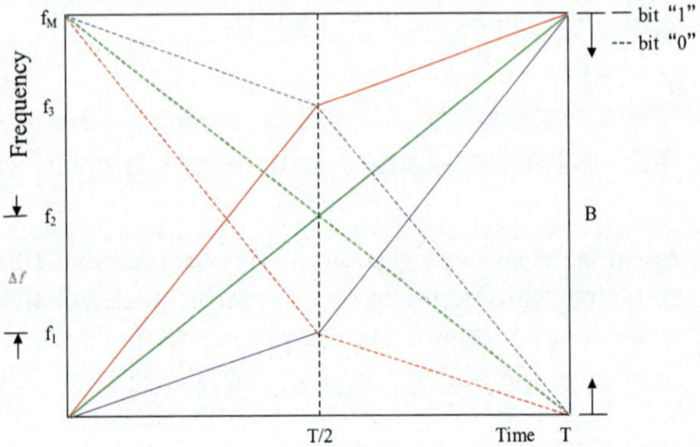

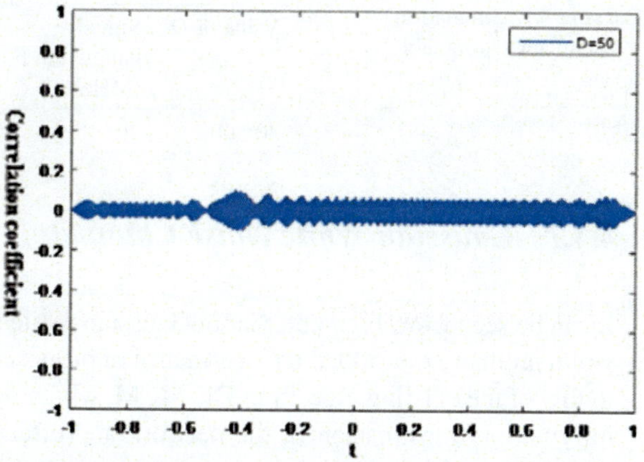


**Fig. 4.4** Cross-correlation coefficient of DT-MCM algorithm

where $k = 1, 2, 3, \ldots, M$ represents the number of users in bee colony network, $M$ is the total number of users, and $E$ is the total energy of the signal in duration $T$. $t_k = \frac{kT}{(k+1)}$ indicates the duration of each chirp rate of each user in network, and $T$ and $B$ are the duration of signal transmission and system bandwidth, respectively. $\mu_k^1$ represents the frequency modulation of the $k$th user in time $0 \le t \le \frac{T}{2}$, and $\mu_k^2$ represents the complementary frequency modulation of the $k$th user in time $0 \le t \le \frac{T}{2}$. The chirp rate of two periods is defined as follows:

$$\mu_k^1 = \frac{(M+1)}{2k}\frac{B}{T} \tag{4.5}$$

$$\mu_k^2 = \frac{(M+1)}{2(M+1-k)}\frac{B}{T} \tag{4.6}$$

Regardless of the interference of non-cooperative parties, the key factor affecting the communication performance of bee colony network on the battlefield is the cross-correlation among network users [9]. Under ideal conditions, a good internet-work multiuser communication algorithm should ensure that the selection signals are orthogonal, that is, the cross-correlation coefficient among users is zero [10, 11]. Figure 4.4 displays the cross-correlation performance of the proposed algorithm drawn by MATLAB. Each signal is approximately orthogonal.

## 4.4 Simulation Results

This section mainly verifies the accuracy of the previous theoretical derivation through MATLAB. Comparing with the existing communication algorithm DF-MCM under Gaussian white noise channel and Doppler channel, proposed DT-MCM's pros and cons are analyzed. The simulation parameter settings are shown in Table 4.1.

**Table 4.1** Simulation parameter setting table

| Value of $D$ | Signal bandwidth $B$ (MHz) | Symbol time $T$ (μs) | Number of simulation data |
|---|---|---|---|
| 50/500 | 20 | 2.5/25 | 400,000 |

### *4.4.1 Gaussian White Noise Channel*

It can be seen from Fig. 4.5a that in Gaussian white noise channel, when time bandwidth product $D = 500$, the performance of proposed DT-MCM algorithm is similar to that of the existing algorithm DF-MCM, which both ensure the reliability of user information transmission in the network and effectively suppress the interference among users. However, while the resources of communication system are limited, that is, when the $D$ value is only 50, as shown in Fig. 4.5b, when $E_b/n_0$ is equal

**Fig. 4.5** Bit error rate curve of the two algorithms under different $D$ values

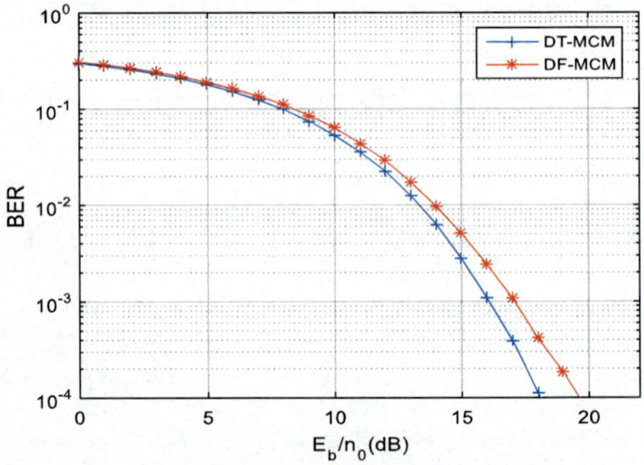


(a) $D = 500$

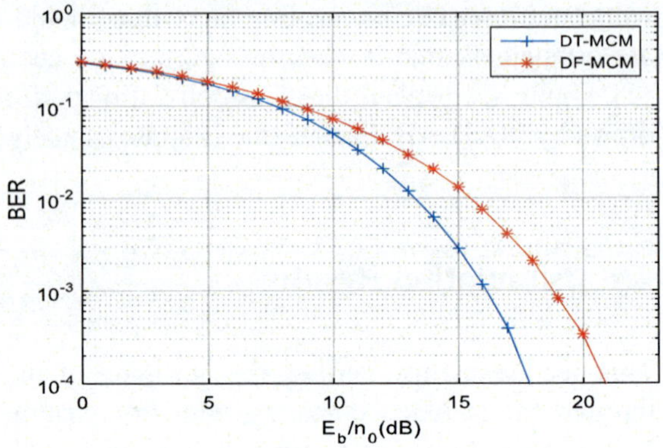


(b) $D = 50$

to 18 dB, proposed DT-MCM theoretical bit error rate of the system can reach at $10^{-4}$, yet the DF-MCM algorithm requires $E_b/n_0$ equal to 22 dB. So, the proposed DT-MCM algorithm has better performance, that is, 4 dB gains.

### 4.4.2 Doppler Channel

In the simulation of air-to-air combat conditions, when there is a Doppler shift in the channel, the performance of two algorithms is simulated by MATLAB. By comparison, the bit error rate curve is shown in Fig. 4.6.

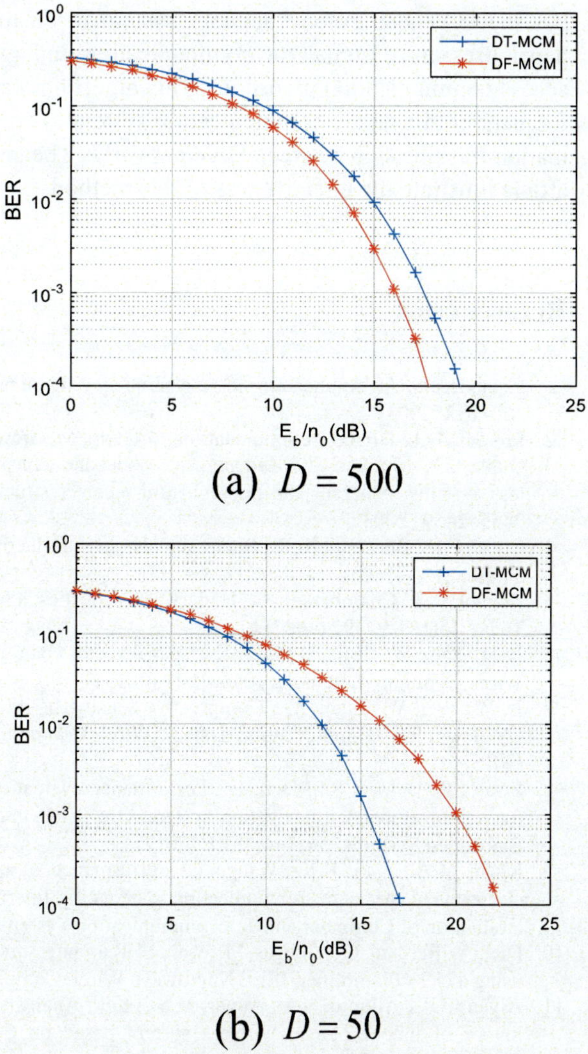


**Fig. 4.6** Bit error rate curves of two algorithms with different $D$ values in Doppler channel

Figure 4.6a demonstrates that when time bandwidth product resources are sufficient, both algorithms have good anti-Doppler capabilities, but at the expense of the same power, the DF-MCM algorithm has better performance. However, when the communication resources are limited, that is, the value of $D$ is greatly reduced, as shown in Fig. 4.6b, the performance of the DF-MCM algorithm is sharply descended, while the DT-MCM algorithm still has a good ability to resist Doppler shift. The performance gain is reduced by about 1 dB. Therefore, the DT-MCM algorithm has better anti-Doppler frequency shift capability.

## 4.5 Conclusions

For the UAV bee colony network system in air-to-air high-dynamic scene, this paper uses chirp signal for communication and gives a new multiuser multiple access communication algorithm. This algorithm can resist the impact of Doppler frequency shift on the communication system to ensure the safe and reliable information transmission between UAV networks. The simulation examples given in the article illustrate the effectiveness of this method.

## References


1. Papandreou-, A.: Applications in time-frequency signal processing. J. Res. Child. Educ. **27**(2), 127–152 (2003)
2. Proakis, J.G.: Digital Communications, 3rd edn. McGraw Hill, Boston (2001)
3. Gupta, C., Mumtaz, T., Zaman, M.: Wideband chirp modulation for FH-CDMA wireless systems: coherent and non-coherent receiver structures. IEEE Int. Conf. Commun. **4**, 2455–2459 (2003)
4. Cook, C.E., Bernfeld, M.R.: Radar signals: an introduction to theory and application. Artech House, MA (1993)
5. Ziomek, L.J., Chamberlain, S.: Underwater acoustics: a linear systems theory approach. Phys. Today **40**(12), 91–92 (1987)
6. Ji, H., Kim, W., Shim, B.: Pilot-less sparse vector coding for short packet transmission. IEEE Wirel. Commun. Lett. **8**(4), 1036–1039 (2019)
7. Ji, W., Wei, W., Liang, Q.: Compressive sensing-based signal compression and recovery in UWB wireless communication system. Wirel. Commun. Mob. Comput. **14**(13), 1266–1275 (2015)
8. Pivato, P., Dalpez, S., Macii, D.: Performance evaluation of chirp spread spectrum ranging for indoor embedded navigation systems. In: IEEE International Symposium on Industrial Embedded Systems, pp. 1–4. IEEE (2012)
9. Khan, M.A., Rao, R.K., Wang, X.: Performance of quadratic and exponential multiuser chirp spread spectrum communication systems. In: International Symposium on Performance Evaluation of Computer and Telecommunication Systems, pp. 58–63. IEEE (2013)
10. Diet, A., Berland, C., Villegas, M., et al.: EER architecture specifications for OFDM transmitter using a class E amplifier. IEEE Microwave Wirel. Comp. Lett. **14**(8), 389–391 (2018)
11. Alyan, E.A., Aljunid, S.A., Anuar, M.S., et al.: Analysis of theoretical and simulated performance of indoor optical wireless system based on CDMA technology. In: International Conference on Electronic Design (2017)

# Chapter 7
# Research on Multi-data Center Collaboration Technology for Multi-station Fusion

**Hu Song, Quan Fang, and Fei Xia**

**Abstract** This paper designs an intelligent collaboration mechanism between the distributed edge data center and the core data center platform. The core data center is responsible for managing and monitoring the distributed data center and the resources of the data center, coordinating the collaborative allocation of resources between the platforms to complete the needs of users. The distributed data center independently manages and monitors the computing resources of the data center and is responsible for processing specific storage and computing tasks. The platforms are connected to each other through the backbone network. This paper innovatively proposes a cloud edge resource graph intelligent scheduling method based on graph database technology that adapts to the collaborative operation of multiple data centers, adapts to multi-center resource collaborative operation scenarios, and is different from traditional centralized data center data resource scheduling methods. The bottom layer is based on graph database technology. Storing resource types and relationships between them, constructing a graph state library describing the correlation between multi-data center cloud-side collaborative resources, and using graph scheduling algorithms to convert heterogeneous multi-data center resource scheduling into directed acyclic graph scheduling problems. Using the minimum cut algorithm of the graph database and the algorithm based on the critical path ensures the optimal scheduling results in a complex multi-data center cloud-side collaborative resource scheduling environment.

## 7.1 Introduction

With the advent of the internet of everything era, the amount of data generated by edge devices has increased rapidly, which has brought higher data transmission bandwidth requirements [1, 2]. At the same time, new applications have also put forward higher requirements for the real-time data processing. The traditional cloud computing model has been unable to respond effectively [3, 4]; edge computing has

H. Song (✉) · Q. Fang · F. Xia
State Grid Jiangsu Electric Power Co., LTD. Infomation and Telecommunication Branch, Nanjing 211106, Jiangsu, China

 
L. C. Jain et al. (eds.), *Smart Communications, Intelligent Algorithms and Interactive Methods*, Smart Innovation, Systems and Technologies 257,
https://doi.org/10.1007/978-981-16-5164-9_7

emerged, and the coordinated development of cloud computing and edge computing will become an important trend in future development [5, 6].

The "multi-site integration" data center refers to the multi-level data center including the edge data center, which is used to store, process and calculate the data generated in the internal business of the power grid, urban governance, internet, industrial manufacturing and other fields. The gathering base of data information is also the main load center. Through the large-scale development of "multi-site integration" data center stations, the overall advantages of edge computing and cloud data centers of "business segmentation, cloud edge collaboration, and overall connection" can be created, which can significantly lead the development of the cloud data upstream and downstream industry chain and promote ubiquitous new business forms and new models of power internet of things appear.

Multi-site integration inevitably has many sites. The grid existing data centers are mostly centralized centers, and the technology accumulation does not cover the coordination problems between multiple distributed centers. It is necessary to study the interaction mechanism and storage technology of resources between core stations and distributed stations to form a resource schedulable whole and exert the synergistic effect of several stations.

## 7.2 Resource Management Architecture of Distributed Data Center and Core Data Center Platform

Based on graph database technology, this paper designs a cloud-side resource intelligent scheduling architecture that adapts to the collaborative operation of multiple data centers. It uses library technology to store heterogeneous resources and their relationships and builds a graph state library describing cloud-side collaborative resources in multiple data centers. The graph scheduling algorithm for multiple data centers achieves optimal scheduling results in a complex heterogeneous multi-data center cloud-side collaborative resource scheduling environment.

The distributed data center and core data center platform resource management architecture is mainly composed of distributed data centers and core data centers deployed in multiple stations, and the platforms are connected to each other through a backbone network, as shown in Fig. 7.1. Difference from single data center, the core data center is responsible for managing and monitoring the distributed data center

**Fig. 7.1** Multi-data center resource management architecture diagram

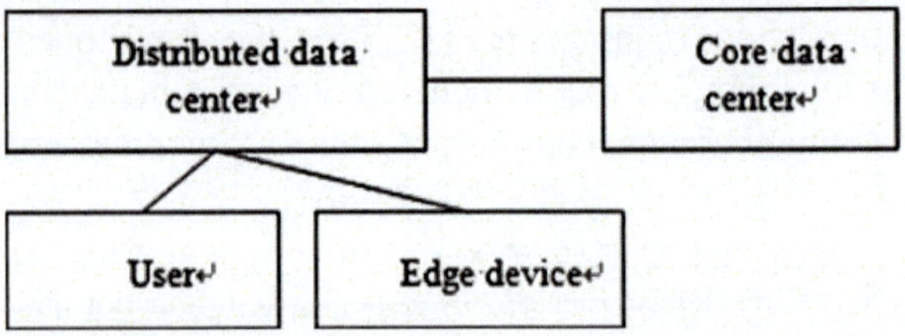

and its own data center resources, and coordinating the collaborative allocation of resources between the platforms to fulfill the needs of users. Each distributed data center independently manages and monitors the computing resources of the data center, and is responsible for processing specific storage and computing tasks.

Different from single data center infrastructure service, this paper proposes a multi data center infrastructure service structure. This architecture mainly provides infrastructure services and meets the requirements for collaborative resource management under a distributed multi-data center architecture. Based on the design concept of distributed data centers, distributed data centers located in different geographical locations can satisfy real-time service requests from local users. The core data center platform and each distributed data center interact with each other through internal and external networks. The isolation mechanism of each distributed data center subnet ensures the secure network isolation of the resources of each data center.

Different from the single data center resource management mode, this paper proposes the core data center platform master control center. The core data center platform master control center is the resource management center of the entire system. It manages and maintains the resource information and data information of all distributed data centers. After the control platforms of each distributed data center are deployed, they can immediately join the overall computing resource pool through registration and access authentication at the master control center. The master control center carries out the coordinated allocation of resources in multiple data centers according to user needs. Each data center is used as a management unit to obtain the resource usage status of each data center; each data center manages each physical server or computing cluster in the area as a unit. It can monitor the operation and load of each physical host in the area.

Users register and authenticate the platform in the core data center through the terminal and can submit resource request tasks to the core data center according to their own needs. The core data center will formulate corresponding strategies and user levels according to user needs, select the appropriate service area and deliver tasks to the corresponding data center and return to the service address of the user data center. At this time, the user can directly access the service provided by the corresponding data center according to the service address, and the terminal will not need to interact with the core data center again during the process of using the cloud service, which increases the real-time nature of the service.

The core data center needs to be connected to each data center. On the one hand, it can obtain the resource usage of each data center in time and synchronize the resource data information in each region in real time in the core data center; on the other hand, it can synchronize each data center. The center issues cloud resource tasks to meet various resource needs raised by users.

## 7.3 Multi-level Allocation and Recycling Technology for Resources in Multiple Data Centers

Based on the resource management architecture, design multi-data center resource allocation and resource optimization recovery plan to realize the optimized management of distributed data center and core data center platform resources. The specific plan to be adopted is as follows in Fig. 7.2.

Under the multi-data center resource management architecture, the core data center formulates resource allocation strategies according to user needs and collaborates to organize the computing and storage resources of each edge cloud. Resource management mainly includes the allocation and recycling of resources. The process of resource allocation in multiple data centers is as follows:

Step 1: The user submits a request for resource application through the platform. The application information includes the area where the resource is located, the type of the resource operating system, and the resource configuration such as the number of CPU cores, memory size, hard disk size, etc.;
Step 2: If the user selects local resources, the corresponding data center can create resource allocation tasks, otherwise the core data center needs to create resource allocation tasks in the master control center;
Step 3: The execution of the task first selects the service area, and then the system forwards the task to the corresponding distributed data center;
Step 4: The resources in the distributed data center adopt the method of polling scheduling to query the running status and resource usage of each physical resource in the data center. If there is no idle resource that meets the request, it will directly return the virtual machine creation failure errors and log records of the system. At the same time, the platform sends the user a notification that the resources in the area are unavailable, and the user re-selects the service area, that is, go to step 1; if an idle resource that meets the request is found, the corresponding physical resource is allocated resource of;
Step 5: If the resource is created successfully, the platform generates the location information of the resource according to the actual situation, including the data center area ID number and resource ID number. Store resource location information and configuration information on the data storage server of the distributed

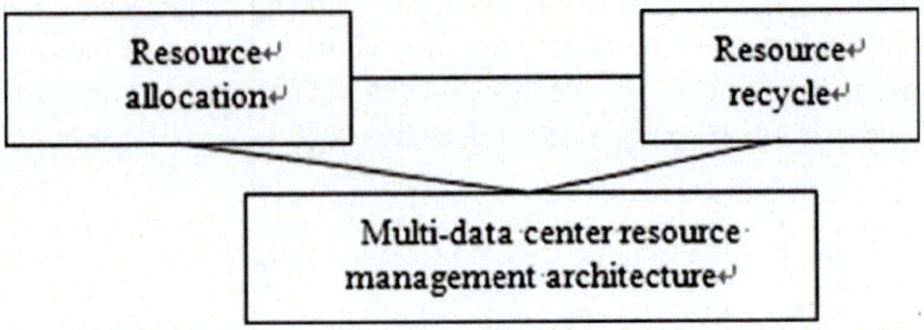


**Fig. 7.2** Virtual and real fusion process diagram

platform and perform data synchronization in the core data center control center in real time, and at the same time, notify the user of the successful resource allocation message and return the resource login information;
Step 6: The resource allocation process ends.

The process of multi-data center resource recovery is as follows:

Step 1: The core data center and the distributed data center obtain the collection of resources that need to be recycled by monitoring the status of resources;

Step 2: Query the location information of the resource to be recycled on the data storage server;

Step 3: The administrator creates a resource recovery task;

Step 4: To execute the task, first select the distributed data center indicated by the ID of the data center area where the resource is located and then forward the task to the corresponding distributed data center;

Step 5: The distributed data center selects the corresponding physical resource or virtual resource in the area according to the server ID;

Step 6: The distributed data center invokes the related methods of the scheduling server to close and recycle resources, and recycle useless resources to the total computing resource pool, and at the same time, update the resource data information in the distributed data center and the master control center and return message of successful resource recovery;

Step 7: The resource recovery process ends.

## 7.4 Data Synchronization Technology for Collaborative Business Between Distributed Data Center and Core Data Center Platform

The design of this paper is based on the multi-data center collaborative difference synchronization method, the design of a custom distributed graph database log format, the application of the data synchronization method based on the update log, and the design of a synchronization transaction plan to achieve transaction-based update and rollback data. In view of the inherent correlation of some update operations, once an error occurs during the synchronization process, simply rolling back or canceling an operation may affect the data in the unpredictable range, resulting in data inconsistency and synchronization failure. And this technology based on research can regard affairs as the basic unit, which improves the security and success rate of synchronization.

The data synchronization method based on the update log includes event trigger and time trigger. In the data synchronization process of the two triggering methods, there may be errors such as data conflicts, transaction blocking, or concurrency. In order to avoid or solve this defect, in the synchronization process, the publishing user and the subscribing user, respectively, adopt the TCP and HTTP transmission protocols, while designing the conflict detection and avoidance mechanism to ensure the integrity and correctness of the transmitted data during the synchronization process.

Synchronous message transmission in distributed systems mainly adopts asynchronous mode and adopts group agent mechanism. The group agent mechanism uses a group of nodes to forward messages and increases the message transmission channels by adding message transmission agents in the network, and improves the reliability of message transmission. This project designs a multi-data center distributed publish/subscribe data transmission model for multi-site converged business rules. This model is oriented to multiple data centers based on a publish/subscribe communication model. Data transmission-related messages are all sent using a group proxy mechanism, making the publishing and subscription of application services more efficient and standardized, and data transmission more secure and reliable. This project is based on the group agent mechanism, combined with business rules, to design an appropriate distributed data center message publishing and subscription mechanism.

## 7.5 Conclusion

This paper innovatively proposes a cloud-side resource graph intelligent scheduling method based on graph database technology that adapts to the collaborative operation of multiple data centers, adapts to multi-center resource collaborative operation scenarios, and is different from traditional centralized data center data resource scheduling methods. The bottom layer is based on graph database technology. Storing resource types and the relationship between them, constructing a graph state library describing the correlation between multi-data center cloud-side collaborative resources, using graph scheduling algorithm to convert heterogeneous multi-data center resource scheduling into directed acyclic graph scheduling problem, using the minimum cut algorithm of the graph database and the algorithm based on the critical path to ensure the optimal scheduling results in a complex multi-data center cloud-side collaborative resource scheduling environment.

**Acknowledgements** This work is supported by Science and Technology Project of State Grid Corporation of China (Research and Application on Multi-Datacenters Cooperation and Intelligent Operation & Maintenance, No. 5700-202018194A-0-0-00).

## References

1. Baker, J., Bond C.: Providing scalable, highly available storage for interactive services. In: Biennial Conference on Innovative Data Systems Research, pp. 223–234 (2011)

2. Armbrust, M., Fox, A.: Griffith, R.: Above the Clouds: A Berkeley View of Cloud Computing. EECS Department, University of California, Berkeley (2009)
3. Manyika, J., Chui, M., Brown, B.: The next frontier for innovation, competition, and productivity, pp. 154–160 (2011)
4. Boettiger, C.: An introduction to Docker for reproducible research. ACM SIGOPS Oper. Syst. Rev. **49**(1), 71–79 (2015)
5. Jo, M., Maksymyuk, T., Strykhalyuk, B.: Device-to-device-based heterogeneous radio access network architecture for mobile cloud computing. IEEE Wirel. Commun. **22**(3), 50–58 (2015)
6. Armbrust, M., Fox, A., Griffith, R., et al.: A view of cloud computing. Commun. ACM, **53**(4), 50–58 (2010)

# Chapter 8
# A Novel Method for Online Control in Equipment Parameters of Blow Molding Process

**Zhisong Zhu, Xu Zhao, and Yu Liu**

**Abstract** A developed method for online control of equipment parameters in blow molding was proposed in this paper. It aimed to investigate the product quality fluctuation caused by the drift of parameters in the production process. The processing parameters and product quality data were got through orthogonal experiments. The product quality index predictor was established by random forest. The product quality indicators were monitored in real time. When the product quality indicators were out of the allowable range, the collaborative control strategy would adjust process parameters to return to the set value according to the quality indicator in time. In this paper, high density poly ethylene (HDPE) was used as the raw material, and the processing of intermediate bulk container (IBC) tank was taken as an example to verify the effectiveness of this method. The experimental results revealed that the developed control method can predict the change of product quality in real time and keep the product quality index to a reasonable range.

## 8.1 Introduction

Hollow blow molding is a molding method that blows a hot-melt parison closed in a mold into a hollow product by means of gas pressure, which is one of the common plastic molding processing methods [1]. The traditional hollow blow molding quality inspection method is manual sampling and measuring the various indexes of the products at intervals. The feedback on the quality of the products has hysteresis, which leads to large change in weight and uneven wall thickness distribution [2, 3] of the hollow blow molding products. The influences of blow molding process parameters on the quality of blow-molded products are complex and nonlinear. The relationship between the blow molding process parameters and the quality of blow-molded products is difficult to be characterized by quantitative formulas. In order to solve the above problems and the fluctuation of some process parameters (process parameter drift) caused by environmental disturbances when producing a tank of

Z. Zhu (✉) · X. Zhao · Y. Liu
School of Mechanical Engineering, Nantong University, Nantong 226019, Jiangsu, China
e-mail: zhu.zhs@ntu.edu.cn

 
L. C. Jain et al. (eds.), *Smart Communications, Intelligent Algorithms and Interactive Methods*, Smart Innovation, Systems and Technologies 257,
https://doi.org/10.1007/978-981-16-5164-9_8

1000 L intermediate bulk container (IBC) [4, 5], this paper proposed a method for online regulation of process parameters based on quality prediction model for which to achieve closed-loop control of product quality.

## 8.2 Product Quality Prediction Mode

In this paper, the orthogonal experiment [6] was used to design 6 horizontal 10 factor orthogonal test; 26 sets of experimental data were studied. The process parameters of blow molding machine [7–9] related to quality index are selected, and the corresponding relationship between process parameters and quality indicators was got. Then a random forest quality model was established to evaluate 26 orthogonal experimental factors. It aimed to eliminate the process parameters which were not related to the quality of the blow-molded products during the blow molding process of plastic products and get 4 levels of 8 factor orthogonal data. The quantitative value of the correlation degree would be the weighting factor of the control parameters in cooperative control. The selected process parameters were training data to establish a random forest quality prediction model.

### 8.2.1 *Orthogonal Design*

The main performance parameters of the blow molding machine included the upper die temperature, the medium die temperature, the lower die temperature [10], melt temperature [11], blowing pressure [12], pressure holding time [13], feeding length, screw speed, and partial temperature of two extruders (18 groups) [14].

Each of the identified test factors determined the respective number of levels based on actual conditions. The principle was to cover all ranges of values for all test factors. For example, the reasonable range of the pressure holding time is 120–135 s. To cover all the values of the test factors and consider the principle of uniform distribution, there were 4 levels to test: 120, 125, 130, and 135 s. According to the actual production conditions, except for the feeding length to be set to 3 levels, the number of other factors is set to 6 levels. The process parameter levels were set as shown in Table 8.1.

### 8.2.2 *Random Forest Prediction Model*

Random forest algorithm is an integrated algorithm [15]. In the model training, the random forest would evaluate the feature relevance to select the most critical process parameters for the quality index. The quantitative value of the correlation degree [16] would be the weighting factor of the control parameter in cooperative control.

**Table 8.1** The process parameters level value

| Influencing factor | 1 | 2 | 3 | 4 | 5 | 6 |
|---|---|---|---|---|---|---|
| Upper die temperature (°C) | 165 | 167 | 169 | 171 | 173 | 175 |
| Medium die temperature (°C) | 165 | 167 | 169 | 171 | 173 | 175 |
| Lower die temperature (°C) | 165 | 167 | 169 | 171 | 173 | 175 |
| Melt temperature (°C) | 170 | 174 | 178 | 182 | 186 | 190 |
| Blowing pressure (Mpa) | 0.54 | 0.56 | 0.58 | 0.60 | 0.62 | 0.64 |
| Pressure holding time (s) | 120 | 122 | 124 | 126 | 128 | 130 |
| Feeding length (cm) | 197 | 198 | 199 | | | |
| Screw speed (Hz) | 40 | 41 | 42 | 43 | 44 | 45 |
| No. 1 extruder temperatures 1–9 (°C) | 150 | 154 | 158 | 162 | 166 | 170 |
| No. 2 extruder temperatures 1–9 (°C) | 150 | 154 | 158 | 162 | 166 | 170 |

According to the correlation analysis, 8 sets of parameters, such as melt temperature, blowing pressure, and feeding length accounted for 90% of the total importance. The feature variables were sorted in descending order of importance, and the attributes that accounts for 10% importance were deleted to get a new feature set. Based on it, a new random forest model was established, and the correlation degree of each feature was calculated in the feature set.

The process parameters were listed according to the correlation degree: the upper die temperature, the medium die temperature, the lower die temperature, melt temperature, blowing pressure, pressure holding time, feeding length, and feeding speed. The test levels for each process parameter were listed in combination with expert experience and actual production as shown in Table 8.2.

**Table 8.2** The main process parameters level value

| Influencing factor | 1 | 2 | 3 | 4 |
|---|---|---|---|---|
| Upper die temperature (°C) | 164 | 168 | 172 | 176 |
| Medium die temperature (°C) | 164 | 168 | 172 | 176 |
| Lower die temperature (°C) | 164 | 168 | 172 | 176 |
| Melt temperature (°C) | 180 | 185 | 190 | 195 |
| Blowing pressure (Mpa) | 0.54 | 0.58 | 0.60 | 0.62 |
| Pressure holding time (s) | 124 | 128 | 132 | 136 |
| Feeding length (cm) | 212 | 214 | 216 | 218 |
| Screw speed (Hz) | 40 | 42 | 44 | 46 |

### 8.2.3 Process Parameter Correlation Analysis

A quality predictor based on random forest regression was established by training experimental data got from orthogonal experiments. Among them, the inputs were the upper die temperature, the middle die temperature and the lower die temperature, melt temperature, blowing pressure, pressure holding time, feeding length, and screw speed. The outputs were the product weight and the wall thickness uniformity of the production (Wall thickness variance).

The parameters of the random forest regression function based on scikit-learn [17] were set as follows:

The maximum number of decision trees: n_estimators = 10;

Evaluation criteria for characteristics: criterion = 'gini';

The maximum depth of the decision tree: max_depth = None;

The minimum number of samples required for internal node subdivision: min_samples_split = 2;

The smallest sample weight of the leaf node is: min_weight_fraction_leaf = 0.0;

Maximum number of leaf nodes: max_leaf_nodes = None;

Whether to put back: bootstrap = True;

The upper die temperatures, the middle die temperature and the lower die temperature were set: $T_{M1} = 172^{\circ}$, $T_{M2} = 168^{\circ}$, $T_{M3} = 168^{\circ}$; the melt temperature $T_R = 187°$; the blowing pressure $P = 0.58$ Mpa; the pressure holding time $t = 130$ s; screw speed $V = 45$ Hz; the feeding length $L = 215$ cm. The test results were shown in Fig. 8.1.

Comparing the error between the test sample values and the quality estimates of the Fig. 8.1a–d, it can be found that the model generalization ability was remarkable. In Fig. 8.1a, the weight of the product decreased with the increase of the melt temperature when other process parameters were unchanged. In Fig. 8.1b, the wall thickness uniformity of the product decreased with the increase of the melt temperature when other process parameters were unchanged. In Fig. 8.1c, the weight of the product was not related to the blowing pressure under the same conditions of other process parameters. In Fig. 8.1d, the variance of the wall thickness uniformity of the product increased with the increase of the blowing pressure under the premise that other process parameters were unchanged. It showed that the quality prediction model got the relationship between the process parameters and product quality indicators.

The random forest algorithm was used to fit the product quality and wall thickness uniformity, respectively. The correlation degree between the product quality and the process parameters was shown in Fig. 8.2. It can be seen that the feeding length, the melt temperature, lower die temperature, and the screw speed played major roles in the weight change of the product. Blowing pressure, melt temperature, lower die temperature, and pressure holding time played a major role in change of wall thickness uniformity correlation degree.

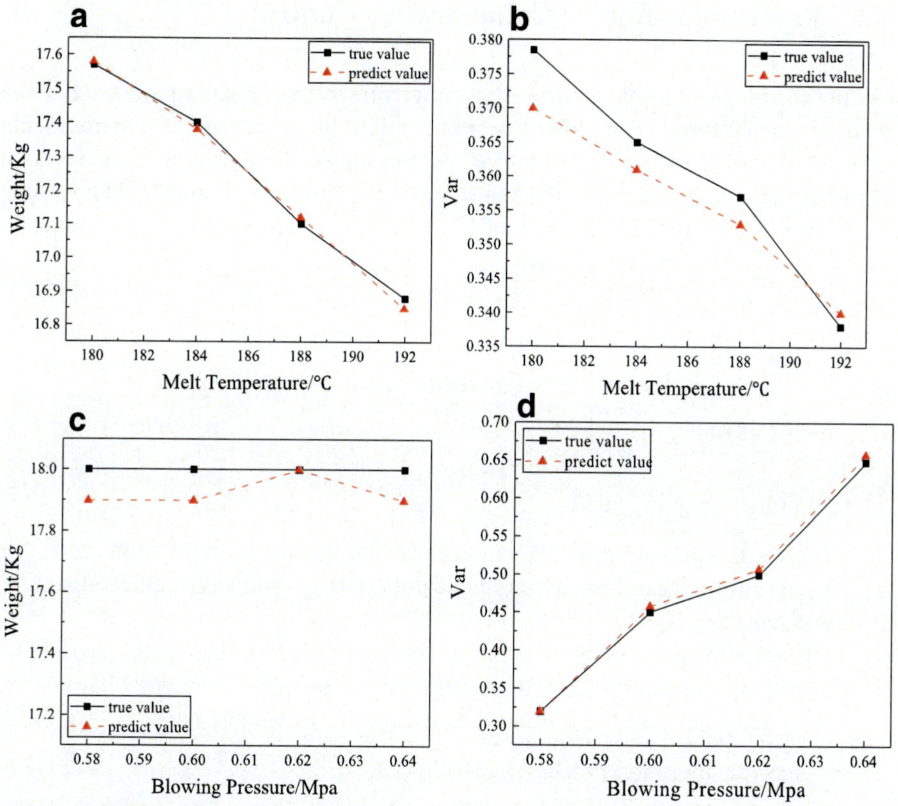


**Fig. 8.1** **a** and **b** were the relationships between the melt temperature and the weight of the product, and the wall thickness uniformity of the production (wall thickness variance); **c** and **d** were the relationships between the lowing pressure and the weight of the product, and the wall thickness uniformity of the production (wall thickness variance)

**Fig. 8.2** The correlation degree between the product quality and the process parameters

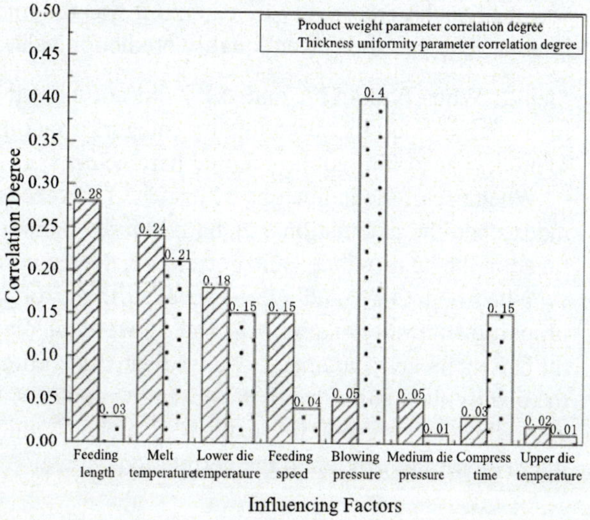

## 8.3 Process Parameter Collaborative Control

During the blow molding process, fluctuations in certain process parameters (such as melt temperature and mold temperature) might be influenced by environmental factors (such as workshop temperature and humidity), which can produce deviation between the actual value and the set value of the process parameter. The resulted deviations would affect product quality; therefore, it was necessary to establish a closed-loop control system to ensure product quality.

### *8.3.1 Process Parameters Collaborative Control Specific Process*

The control strategy included:

1. The current process parameters were put into the quality prediction model, the model output the current quality indicators, and determined whether adjustment was needed;
2. The relevant process parameters were adjusted according to the degree and direction of the correlation parameters of the process parameters. Ultimately, the quality index of the product was within the acceptable range.

The specific operation process was followed:

Step 1: The weight index predicting the quality fluctuation value $\Delta Y_{1\mathrm{i}}$ and the wall thickness uniformity index predicting the quality fluctuation value $\Delta Y_{2\mathrm{i}}$ were calculated, the current parameters were put into the random forest quality model to predict the quality index, and judged whether it exceeds the index range by $|\Delta Y_{ki}| > e_k$, $e_k$ was the quality index error range. The value of $\Delta Y_{\mathrm{ki}}$ is $\Delta Y_{ki} = Qs_k - Q_{ki}$, $Qs_k$ was the benchmark quality index, and it was the middle value of the quality index interval. $Q_{ki}$ was the current quality prediction value.

Step 2: When both $\Delta Y_{1i}$ and $\Delta Y_{2i}$ did not exceed the error range of the quality index, it indicated that both quality categories met the quality index and continued to process according to the original parameters.

When one of the indicators $\Delta Y_{1\mathrm{i}}$ and $\Delta Y_{2\mathrm{i}}$ exceeded the error range of the quality index, then the calculation was based on the parameters which exceeded the error range of the quality index and correlation degree corresponding to the indicators.

When both $\Delta Y_{1i}$ and $\Delta Y_{2i}$ exceeded the error range of the quality index, the proportion of limit violations $\beta_{1i}$ and $\beta_{2i}$ were calculated and compared. If $\beta_{1i} > \beta_{2i}$, the current process parameters were adjusted according to the weight index predicting the quality fluctuation value $\Delta Y_{1i}$. On the contrary, the current process parameters were adjusted corresponding to the wall thickness uniformity index predicting quality fluctuation value $\Delta Y_{2i}$ and the correlation degree.

Step 3: Adjust the current process parameters.

The adjustment process is as followed:

Calculate the adjustment amount of each iteration $U_{ij}$:

$$U_{kij} = A_{kj} \times B_{ij} \times \beta_{ki}(1 + \alpha_{ki}) \quad (8.1)$$

$A_{kj}$ was the process parameter correlation quantified value, $B_{ij}$ was the current process parameter,

$\beta_{ki}$ was the overrun ratio, its value was:

$$\beta_{ki} = \Delta Y_{ki}/Qs_k \quad (8.2)$$

$\Delta Y_{ki}$ was the current predicted quality fluctuation value, $Qs_k$ was Benchmark quality indicator.

$\alpha_{ki}$ was the adjustment speed, the size was:

$$\alpha_{kij} = \left[\Delta Y_{ki} - \Delta Y_{k(i-1)}\right]/\Delta Y_{ki} \quad (8.3)$$

Calculate the adjusted process parameters $B_{(i+1)j}$:

$$B_{(i+1)j} = B_{ij} + U_{kij} \quad (8.4)$$

Determine if $B_{(i+1)j}$ was outside the process parameter range $\left[B_{j\,\min}, B_{j\,\max}\right]$. If it was not exceeded, the process parameters were in line with the quality index; if it is exceeded the range, the current process parameter category $B_{ij}$ was unchanged. After the current process parameter category was removed, the remaining process parameter correlation quantified value $A_{kj}$ was recalculated, $U_{kij}$ and $B_{(i+1)j}$ were recalculated;

Return to step 1 and iterate until the quality indicator was met.

In the above process, $k$ was the quality indicator category, $j$ was the process parameter category, and $i$ was the iteration count.

The specific adjustment process was shown in Fig. 8.3:

### 8.3.2 Experimental Verification

In order to verify the effectiveness of the process parameter collaborative control method, the melt temperature reduced by 8°to simulate a change in the situation of environmental. The data of product weight and wall thickness uniformity in each production cycle were recorded separately, and the change of product quality and wall thickness uniformity under coordinated control and without intervention were compared. The result was shown in Fig. 8.4.

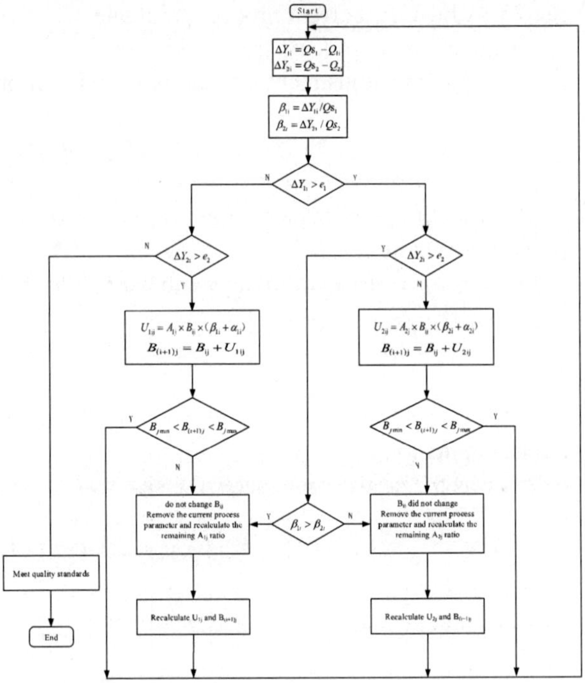

Fig. 8.3 A schematic diagram of the adjustment process

By comparison, the collaborative control method can find the combination of process parameters by the coordinated adjustment of process parameters, so the product quality can return to the qualified range and avoid the situation of a large number of unqualified products. The quality parameters after cooperative control had been in a reasonable range and the fluctuations were small.

## 8.4 Conclusion

In order to solve the problem of process parameter drift in the actual production process of 1000 L intermediate bulk container (IBC) tank, this paper proposed an online regulation method for process parameters of blow molding process. The method monitored the quality of the product in time by establishing a random forest quality prediction model. When the product quality fluctuated too much, the process parameters of the product would be regulated by the collaborative control method indirectly. The results showed that the control method can monitor the quality changes of the products and restore the quality to a reasonable range in time. The method reduced process parameter drift and provided certain references for the online quality control of blow-molded products.

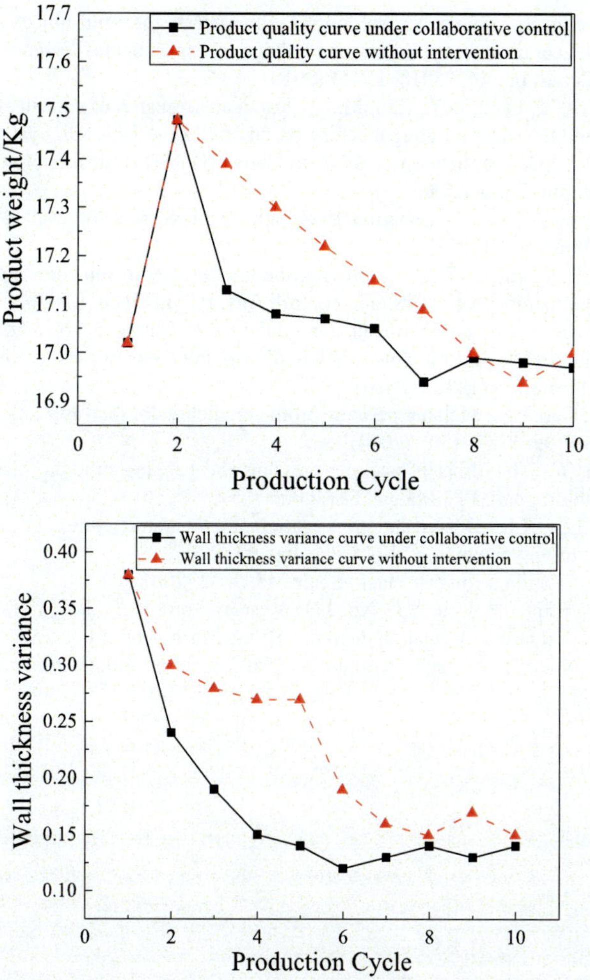


**Fig. 8.4** The data of product weight and wall thickness uniformity in each production cycle were recorded separately, and the change of product quality and wall thickness uniformity under coordinated control and without intervention were compared

## References

1. Au, K.M., Yu, K.M.: Conformal cooling channel design and CAE simulation for rapid blow mould. Int. J. Adv. Manuf. Technol. **6**, 311–324 (2013)
2. Zeng J.S., Zeng, X.F.: Research on Wall Thickness Control of Parison by Hollow Blow Molding. China Plastics Ind. 09 (2010)
3. Musa, R.K, Tan, V., Kalyon, D.: Measurement and calculation of parison dimensions and bottle thickness distribution during blow molding. Polym. Eng. Sci. **21**, 331–338 (1981)
4. Glor, M.: Requirements for flexible intermediate bulk containers (FIBC) current status of standardization. Sic Handhabung Brennbarer Sta. **2182**, 163–174 (2012)
5. Nayan, P., Sunny, Zr., Himanshu, P.: Performance evaluation of HDPE/MWCNT and HDPE/kenaf composites. J Thermoplast. Compos. (2019.08)

6. Deng, L., Feng, B.., Zhang, Y.: An optimization method for multi-objective and multi-factor designing of aceramic slurry. Combining orthogonal experimental design with artificial neural networks. Ceram Int. **44**, 15918–15923 (2018)
7. Li, W.M., Xu, Y., Feng, D.Y., Zho, M.D.: Design and research of control system of the large-scale accumulator blow molding machine based on PLC. Adv. Intell. Syst. Res. **23** (2012)
8. Wang, A., Li, X.J.: Application of BP neural networks PID control in blow molding machine system. Comput. Simula. **09** (2010)
9. Wang, L., Gao, Y., Xi, Z.: Application of computer technology in hollow blow molding machine. Plast. Sci. Technol. **10** (2015)
10. Ruan, Y.F., Gao, H., Li, D.Q.: Improving the consistency of injection molding products by intelligent temperature compensation control. Adv. Polym. Tech. (2019)
11. Biglione, J., Bereaux, Y., Charmeau, J.Y,. Balcaen, J., Chhay, S.: Numerical simulation and optimization of the injection blow molding of polypropylene bottles—a single stage process. Int. Mater. Form. **9**, 471–487 (2016)
12. Zhao, C.H., Gao, F.R.: Melt temperature profile prediction for thermoplastic injection molding. Polyneng Sci. **39**, 1787–1801 (2004)
13. Qiu, B., Li, W.: The effect of packing pressure and packing time on thickness distribute of injection moldingpart. Pla. Manuf. **11** (2006)
14. Du, Q.L., Chen, X.H.: Design and investigation of temperature control system for hollow blow molding extruding machine. CHINA Plas Ind. **07** (2006)
15. Breiman, L.: Random forests. Mach. Learn. **45**, 5–32 (2001)
16. Huang, F.L., Lin, J.P., Lou, Y.J., Xu, J.H.: Research on robust design of craft parameters in plastic injection based on relation degree. CHINA Mech. Eng. **21** (2009)
17. Wang, Y.S., Xia, S.T.: A survey of random forests algorithms. Inf. Commun. Technol. 01 (2018)

# Chapter 9
# Research on the Optimum Ratio of Intelligent Rail Guided Vehicle and Computer Numerical Control

**Man Li, Xuewei Huang, Jiaqi Wang, Haiyun Yue, and Chaolang Hu**

**Abstract** In the intelligent machining system, the ratio between the number of track-type automatic guide vehicle and the number of computer numerical control (abbreviation CNC will be used below) machine tool, as well as its scheduling plan have a decisive influence on the machining efficiency of the system. In this paper, we explore the optimum ratio of rail guided vehicle (abbreviation RGV will be used below) and CNC corresponding to the processing time of different individual parts under single process. Through dividing the processing time of a single part, the "2n step by step" model is designed for the RGV. And the traversal algorithm is used to find out the optimal path planning of RGV for all the CNC values in a given processing time. Then, by comparing the total waiting time of RGV and CNC in different cases, the optimum ratio of the certain processing time is determined, which provides a strong basis for the manufacturer on how to design start-up plan.

## 9.1 Introduction

With the development of information technology, big data analysis and artificial intelligence technology, more and more problems of optimal scheduling and optimal control have emerged in many fields, such as industrial manufacturing, machining and logistics transportation. The problems have a wide range, such as material processing scheduling, job process scheduling, warehouse material storage scheduling, classification of logistics products and so on. RGV is a kind of unmanned intelligent car, which can run freely on fixed track. It can do simple repetitive work, reduce the waste of human resources and effectively improve the production efficiency of modern intelligent processing factory [1]. Thus, it is widely used in workshop and automatic three-dimensional warehouse. Its working range includes goods storaging, goods picking, AB conversion and other tasks. Particularly, RGV can cooperate with CNC to participate in loading, unloading and cleaning of a variety of parts that require different processing times. At this time, RGV and CNC form a processing system.

M. Li · X. Huang · J. Wang · H. Yue · C. Hu (✉)
Math College of Sichuan University, Chengdu, China
e-mail: huchaolang@scu.edu.cn

 
L. C. Jain et al. (eds.), *Smart Communications, Intelligent Algorithms and Interactive Methods*, Smart Innovation, Systems and Technologies 257,
https://doi.org/10.1007/978-981-16-5164-9_9

Obviously, the working efficiency of the system is closely related to the number of RGV on track and the number of CNC, and the route planning of the system is also an important factor.

Up to now, some articles have already studied the related problems of RGV. Zhang Guiqin and others [2] used collision avoidance algorithm to avoid the problem of two RGV colliding or trailing in the same direction; aiming at the influence on control and operation performance caused by time-varying uncertainty of sampling period in network control system. Shen Yan and others [3] proposed dynamic scheduling strategy based on feedback control principle and prediction mechanism. On this basis, Cha Zhenyuan and others [4] compared the usage of RGV and stacker under six scheduling strategies and determined the scheduling strategy with the highest utilization.

Nevertheless, as mentioned above, the existing research results mainly focus on route design, and the ratio of RGV and CNC is mostly processed by a fixed value. But the optimal ratio between intelligent rail guide vehicle and CNC machine tool is seldom studied, and there is a lack of more general and universal conclusions. In fact, when there is a serious mismatch of the ratio between RGV and CNC, no matter how the scheduling scheme is designed, the production efficiency cannot be greatly improved. Because the RGV processing time and waiting time will become the bottleneck of the system. So we think when the ratio is fixed, the study on strategy scheduling has practical significance, and under the specified scheduling algorithm, according to the relative proportion relationship between the RGV processing time of a single product and the CNC processing time of a single product, studying the optimum proportion of RGV and CNC is equally important. Our research conclusions can provide a reference for manufacturers to determine how many CNC to start at the same time according to the CNC processing time of a single product, which will help to avoid the second change caused by unreasonable distribution of RGV and CNC quantity, as well as to save time and cost.

Therefore, based on the CNC processing time of a single part, this paper establishes a RGV path planning model for each type of data, and optimizes the scheduling with the shortest sum of CNC waiting time in one shift (8 h) as the goal. Moreover, after the optimization scheme is obtained, the CNC optimal proportion determination model is established to find the optimal proportion and the corresponding scheduling scheme by minimizing the sum of the waiting time of RGV and CNC under the optimization scheme corresponding to each CNC value. The summary results provide the theoretical basis for the proportion setting in production practice, and at the same time, general conclusions are drawn, and the next research direction is explained.

## 9.2 Problem Analysis

The working environment studied in this paper consists of a limited number of CNC, 1 intelligent rail guide vehicle, 1 straight track, 1 feeding conveyor belt, 1 blanking conveyor belt and other related auxiliary facilities. We stipulate that the CNCs are always placed by this rule: odd number on one side, and the corresponding even number on the other side. For example, when there are 8 CNC, the whole system is shown in the Fig. 9.1.

Under the above working environment, each CNC is equipped with the same cutting tools for the same processing process. Each raw material must be processed on any CNC and cleaned by the RGV before it can be finished. This paper always considers the number of RGV to be 1 when discussing the optimal ratio of RGV and CNC corresponding to different processing time T, that is, only the number of CNC is changed. Since between the opposite two CNC, RGV does not need to spend extra moving time on loading material; it is considered that the actual production situation is more favorable when CNC appear in pairs. So in the follow-up discussion, the selection of CNC quantity is all in pairs. For avoiding unnecessary calculations and making the results more general, we first divide the processing time T into intervals. Since too long processing time is meaningless to the production mode discussed in this paper, combined with the actual production data, this paper discusses the processing time within 0–1200 s.

First, the scheduling strategy of RGV with a fixed ratio of 1:$n$ is discussed. In the process of optimizing the ratio, our ultimate goal is always to improve the processing efficiency of the system, that is, we hope that both RGV and CNC are in the working state as long as possible. Therefore, in a given processing time interval, we discuss the optimal path planning for RGV and then compare the sum of waiting time to select the optimum ratio corresponding to the processing time.

When it comes to the problem of the optimal path planning, considering that when all the CNC are in the processing state, if RGV waited and not going anywhere after it completes the current instruction, unnecessary waiting time will be generated before any CNC sends signal to the RGV again, which will directly lead to the reduction of production efficiency. Therefore, when scheduling the RGV, it is stipulated that if the RGV does not receive any signal of feeding demand after completing the current instruction, the RGV will immediately discriminate according to the scheduling model and execute the moving instruction. As far as possible to increase the efficiency of the operation, this paper hopes to increase the number of finished materials in a shift, which means to make the CNC stay in work state as long as possible. In other words, the total waiting time of the CNC should be as small as possible. Since only local optimal solutions can be obtained for such problems, we consider designing a "2n step-by-step" scheduling model for RGV, where n is half of the number of CNC opened in the current planning. RGV will traverse $2n$ times to calculate the future movement (or stay-in-place) before discriminating the execution of the moving instruction, and it is required that each movement or stay-in-place must be separated by more than one instruction to move or stay here. After traversing the future movement for $2n$ times, RGV will take the first step of the path with the minimum total waiting time (or stay-in-place) as its next instruction.

Based on the above calculation, when n changes from 1 to 10, according to the comparison of the total waiting time of RGV and CNC, we determine the optimal ratio of RGV and CNC for each case.

## 9.3 Model Building

### 9.3.1 Determination of RGV Movement Time

Suppose that RGV moves at a constant speed, the time required to move 1, 2 and 3 units of RGV is obtained according to the data consulted. Further, the time required to move 0–10 units of RGV motion is obtained by fitting with the least square method. The fitting function is $f = 16.6x + 1.6$.

### 9.3.2 Estimation of CNC Upper Limit

When RGV movement time is neglected ideally, the production of a finished material requires at least RGV loading, unloading and cleaning and one CNC processing. Since the loading and unloading time and cleaning time generally do not fluctuate greatly, and there is no substantial impact on the discussion of this problem, combined with the existing data, the cleaning time is taken as 27 s, numbered odd CNC the loading and unloading time is 28 s, numbered even CNC the loading and unloading time is 33 s. Therefore, the sum of loading, unloading and cleaning time for one product is 55 or 60 s. It is supposed that the total duration of a shift is $TT = 8\,\text{h} = 28800\,\text{s}$.

So in the case of short processing time, RGV could process up to 525 finished materials within a shift. Accordingly, the CNC upper limit should be

$$\left\lceil \frac{1}{2} * \frac{525}{\frac{18800}{1200}} \right\rceil = 10(\text{pairs}) \tag{9.1}$$

Thus, for each range of processing time, we only consider the number of CNC between 1 and 10 pairs.

### 9.3.3 A Preliminary Division of Processing Time T for a Single Material

Assuming that for a single process, the time of CNC machining single material is constant and consistent. Since too long processing time is meaningless to the production mode discussed in this paper, combined with the actual production data, this paper discusses the processing time within 0–1200 s. According to the data table of operation parameters of intelligent processing system, the processing time of CNC of a single material varies greatly under different specific conditions. To consider the problem comprehensively as far as possible under the condition of reducing the complexity, the partition interval is taken for 10 s. This kind of division is more representative and has strong operability in production practice.

### 9.3.4 The Optimal Path Planning Model

Working principle of CNC. Assuming that in one shift (8 h), CNC working state is stable, that is, failure is not considered. There are four states of CNC: idle state, loading and unloading state, processing state and demand state in the whole course of working. The idle state represents the state of no material on the CNC; the loading and unloading state represents the state of the RGV is feeding the CNC; the processing state represents the state of the material being processed; and the demand state represents the state of waiting for the RGV blanking after the completion of the processing.

Let the state of the CNC numbered $i$ be $A_i$, and assign values to each state as follows.

$$A_i = \begin{cases} 0, \text{idle state} \\ 1, \text{loading and unloading state} \\ 2, \text{processing state} \\ 3, \text{demand state} \end{cases} \tag{9.2}$$

where the values 0 and 3 indicate that the CNC is in a non-working state; according to the goal of improving production efficiency, we naturally hope that the time of these two states is as short as possible.

Working principle of RGV. Assuming that the conveyor belt can deliver the material to the feeding place in time, that is, there is no waiting time for the conveyor belt. Then in the whole course of working, the RGV is always located on the straight track, there are four states: waiting state, moving state, loading and unloading state and cleaning state. The waiting state indicates the state that RGV is resting on the track with no work; the moving state indicates the state when RGV is driven from one position to another; the loading and unloading state indicates the state in which RGV is feeding the CNC; and the cleaning state indicates the state when RGV is cleaning the material and placing it to the conveyor. RGV can operate in strict accordance with the operating instructions [5], set the state of the RGV as $B$ and assign values to each state as follows.

$$B = \begin{cases} 0, \text{waiting state} \\ 1, \text{moving state} \\ 2, \text{loading and unloading state} \\ 3, \text{cleaning state} \end{cases} \tag{9.3}$$

where 0, 2, 3 states mean that RGV can only be in a fixed position on the track, at this time it can only be linked with the adjacent CNC of the position. Let $j$ be where the RGV is, then

$$j = \begin{cases} 1, i = 1 \text{ or } 2 \\ 2, \ i = 3 \text{ or } 4 \\ \ldots\ldots \\ 10, \ i = 19 \text{ or } 20 \end{cases} \tag{9.4}$$

When the RGV is in 1 state, let its motion state be $(j_1, j_2)$, indicating that RGV is moving from position $j_1$ to position $j_2$ at this time.

When the material is in the processing state, the corresponding CNC is in the working state. The goal of the planning is to improve the processing efficiency, that is, the more materials processed in the same time. Obviously, the shorter the non-working time of the CNC, the higher the processing efficiency.

Determination of the objective function. Considering the shortest time for the CNC to be in a non-working state in one shift. For a complete machining process, the machining system starts at 0 s, RGV executes instructions from the initial position and at 28,800 s the machining system stops.

For a given processing time, let n pairs of CNC be turned on ($n$ = 1, 2, ... 10) Define the total waiting time for CNC as follow.

$$H_I = \sum_{i=1}^{2n} \sum_{k=1} h_{ik} \tag{9.5}$$

where $H_I$ is the total waiting time of CNC in the $I$th case, $h_{ik}$ is the waiting time in the $k$ operation of the CNC numbered $i$, and the material number is recorded according to the order in which the raw material is grabbed.

By formula (9.5), in order to minimize the total waiting time of the CNC, the objective function is defined as the following optimization function.

$$\min H_I \tag{9.6}$$

Initial scheme planning. Step 1: The initial case is considered firstly. The initial position of the RGV is position 1, and the CNC are all idle. The initial instruction is to feed those $2n$ CNC in certain order.

Step 2: When the RGV is in a certain position, loading for the two adjacent CNC in that position does not need extra moving time. So when RGV is loading for one CNC at a certain location, the next loading target gives priority to another CNC at that location.

Step 3: RGV first feeds the two CNC at position 1, then there are $(n - 1)!$ possible initial feeding orders. Thus, the equivalent transformation of the objective function is as follows.

$$\min\{H_{I1}, H_{I2}, \ldots, H_{I(n-1)!}\} \tag{9.7}$$

Optimization model of scheduling scheme. Step 1: Taking the next move of RGV into account after finishing the feeding of $2n$ CNC for any initial plan. Although the cleaning step is added in the subsequent processing, the complete round of loading and unloading for $2n$ CNC will be given priority in the actual processing. So the dynamic planning of $2n$ consecutive complete operations is considered for RGV.

Step 2: Since RGV is in a certain position, loading and unloading for the two CNC in that position do not need extra moving time, so it is considered that the next loading and unloading target of the RGV should give priority to the other CNC directly opposite the current one. That is, RGV will give priority to the two CNC at the same position in turn. Therefore, for the next $2n$ jobs, there are $n^{2n}$ types of scheduling schemes.

Step 3: Assume that RGV has memory storage function and can record the time of last interaction with each CNC. For any scheduling scheme, each operation of RGV will result in a waiting time $h_{im}$ for the CNC numbered $i$, where $m$ represents the $m$th of the $2n$ consecutive jobs. For this scheduling scheme, the total CNC waiting time is as follows.

$$h_l = \sum_{m=1}^{2n}\sum_{i=1}^{2n} h_{im} \tag{9.8}$$

The $l$ here represents the $l$th $2n$ consecutive job in the whole process.

Step 4: Find out the total waiting time under each scheduling scheme; there are $n^{2n}$ results. Take the scheme corresponding to the result with minimum total waiting time, that is, minimize the formula (9.8). The corresponding objective function are as follows.

$$\min h_l = \sum_{i=1}^{2n} h_{il} \tag{9.9}$$

Step 5: Choose the first step of the $2n$ consecutive jobs chosen above for RGV to execute. Then all CNC in this step have a total waiting time as follows.

$$hh_l = \sum_{i=1}^{2n} h_{il} \tag{9.10}$$

Take the end of the step chosen above as the new beginning, repeat step 3 and step 4 and perform a number of $2n$ consecutive jobs until the end of a shift.

For the $v$th initial case ($v \in \{1, 2, \ldots, (n-1)!\}$), the total waiting time for CNC resulting from the corresponding complete scheduling scheme is as follows.

$$H_{Iv} = \sum_{l=1} hh_l \tag{9.11}$$

Step 6: Synthesize step 1 to step 5 and minimize the formula (9.11); the complete optimization model is as follows.

$$\min\{H_{Iv}\}$$
$$\text{s.t}\begin{cases} H_{Iv} = \sum\limits_{l=1} hh_l \\ hh_l = \sum\limits_{i=1}^{2n} h_{il} \\ \min h_l = \sum\limits_{m=1}^{2n} \sum\limits_{i=1}^{2n} h_{im} \end{cases} \tag{9.12}$$

### 9.3.5 Determination of Optimum Ratio of CNC

On the basis of practical experience, when the ratio of RGV and CNC is appropriate, they will not have to wait long for each other. So we consider choosing the shortest sum of total waiting time between RGV and CNC as the optimization goal. The optimal path planning model shows that when the number of the open CNC is $2n$, the total waiting time $H_{In}$ of the CNC is also determined after the optimal path is determined. Let the corresponding number of finished materials be $N_{In}$.

To RGV, the assumption above shows that RGV won't wait and not going anywhere after a cleaning task but immediately make a judgment and starts moving. Assume that the RGV works well, it moves strictly according to the instructions. When the RGV moves in front of the selected CNC, but the processing of this CNC is not completed, the waiting time of RGV will generate. Introduce the 0–1 variable $D_{ik}$ as follows.

$$D_{ik} = \begin{cases} 1, & \text{the } k\text{-th feeding occur in the } i\text{-th CNC} \\ 0, & \text{else} \end{cases} \tag{9.13}$$

During the $k$th feeding, the RGV arrives at the selected CNC at time $t_{1k}$, and the CNC completes the current material processing at time $t_{2k}$, introduce the 0–1 variable $\delta_k$ as follows.

$$\delta_k = \begin{cases} 1, & t_{2k} - t_{1k} > 0 \\ 0, & t_{2k} - t_{1k} \leq 0 \end{cases} \tag{9.14}$$

Let $Q_{In}$ be the waiting time of RGV in a shift under the $I$th case with $2n$ open CNC. Then

$$Q_{In} = \sum_{k=1}^{N_{Ik}} \sum_{i=1}^{2n} D_{ik}\delta_k (t_{2k} - t_{1k}) \tag{9.15}$$

In summary, the objective function is as follows.

$$\min_n \{H_{In} + Q_{In}\}, n \in \{2, 3, \ldots, 10\} \tag{9.16}$$

## 9.4 Solution and Result

### *9.4.1 Solution of the Optimal Path Planning Model*

From the above, the set of processing time can be represented as follows.

$$A = \{T | T \in [10, 1200] \text{ and } T|10\} \tag{9.17}$$

For each case ***I*** that the processing time $t \in \mathcal{A}$, consider the number of CNC from 1 to 10, respectively, to different scheduling scheme using traversal algorithm in MATLAB to solve. However, considering the amount of calculation and time, "five step by step" is used in the optimization model, which means in the initial state we get $2n$ steps as a process, then search the optimal path of five operations. After getting the processing scheme for the minimum total waiting time, compared the sum of the waiting time of RGV and each CNC corresponding to the optimal scheduling scheme for the different value of CNC number, and the optimal CNC number of the value of each machining time and the corresponding CNC working efficiency are finally obtained. Suppose that under the case $I$, the ideal number of finished materials is as follows.

$$N_I = 2m_I \left[ \frac{TT - 2m_I s}{T + s} \right] \tag{9.18}$$

where $[\cdot]$ represents round down. Suppose that under the case $I$, the actual number of finished materials corresponding to $n$ pairs of CNC is $N_{In}$, then there is a corresponding rate of finished materials.

$$\alpha_{Ik} = \frac{N_{Ik}}{N_I} \times 100\% \tag{9.19}$$

and the maximum number of materials processed.

### 9.4.2 *The Solution of Optimal CNC Number Ratio*

After the actual calculation, the sum of the processing time of RGV and each CNC is taken as the judgment index to balance the quantity of finished materials, the working efficiency of $\alpha_{Ik}$ and the quantity cost of CNC, so as to ensure the objectivity of the model and the rationality of the results.

The result using traversal algorithm in MATLAB is shown in Fig. 9.2.

According to the calculation results, when the processing time $T \in [10\text{ s}, 210\text{ s}]$, the optimal number of CNC pairs is 1. When the processing time $T \in [220\text{ s}, 360\text{ s}]$, the optimal number of CNC pairs is 2. When the processing time $T \in [370\text{ s}, 510\text{ s}]$, the optimal number of CNC pairs is 3. When the processing time $T \in [520\text{ s}, 650\text{ s}]$, the optimal number of CNC pairs is 4. When the processing time $T \in [660\text{ s}, 800\text{ s}]$,

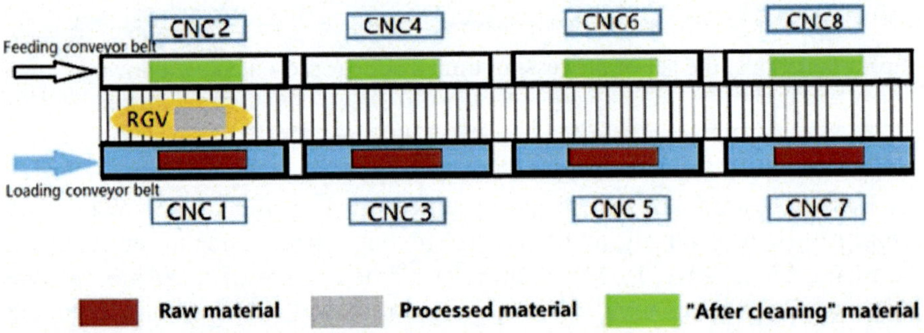


**Fig. 9.1** Schematic diagram of intelligent processing system ( adapted from the "2018 Higher Education Cup National Mathematical Modeling Competition for College Students" problem B: intelligent RGV dynamic scheduling strategy)

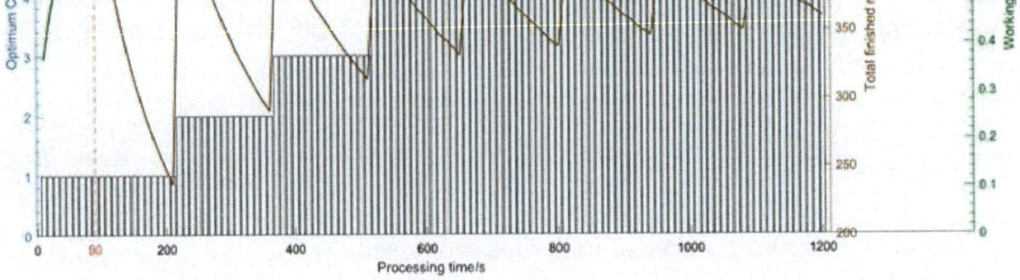


**Fig. 9.2** The optimum CNC quantity, working efficiency and total material number corresponding to the processing time of CNC single part

the optimal number of CNC pairs is 5. When the processing time $T \in [810\ \text{s}, 940\ \text{s}]$, the optimal quantity of CNC pairs is 6. When the processing time $T \in [950\ \text{s}, 1080\ \text{s}]$, the optimal quantity of CNC pairs is 7. When the processing time $T \in [1090\ \text{s}, 1200\ \text{s}]$, the optimal number of CNC pairs is 8.

## 9.5 Conclusions

In this paper, the optimal path of RGV is planned by considering the future steps to determine the current step to approach the global optimal solution, which avoids complex global search and random generation with strong uncertainty, but still obtains the ideal result. When the optimal number of CNC is determined, this paper considers both processing efficiency and resource utilization, which is closely related to the actual production situation and has strong operability. According to the solution results, a general conclusion can be drawn: as the value of processing time increases, the optimal number of CNC increases. Meanwhile, from the calculation results, expect the results of poor preferability with the extreme value of the processing time $T \in [10\ \text{s}, 90\ \text{s}]$, it can be found that the corresponding processing time amplitude of each optimal CNC number fluctuates within 110–140 s, and it is speculated that the corresponding relation has a certain periodicity. Whether this property is affected by the value of the time taken by RGV for CNC to load and unload materials, and the time taken by RGV to clean materials needs further research and explanation.

This paper not only gives the optimal quantity ratio of CNC for different processing times, but also gives the corresponding scheduling scheme, which has a profound practical significance for production. However, when the processing time is shorter, namely less than 90 s, due to the large deviation in measuring the processing efficiency, the operability is not strong; hence, separate consideration should be given to the ratio problem in this processing time interval. In addition, the speed of the conveyor belt in actual production will also affect the scheduling scheme. In future studies, the influence of random faults can be considered, and the fault time and repair time can be simulated by Monte Carlo method. On this basis, the path planning of RGV can be carried out and the optimal ratio can be found.

To sum up, this paper mainly uses dynamic programming to design the optimal scheduling scheme for the intelligent machining system and carries out the optimal ratio of the quantity of CNC and RGV according to different processing time. In this paper, the methods of "multi-step determination in the future of one step" and "screening based on processing efficiency and resource utilization" can be extended to more scheduling and matching problems so as to discuss the optimal grouping of CNC and RGV when the number of RGV increases.

## References

1. Cai, X.D.: 2020 International Conference on Artificial Intelligence and Communication Technology (AICT 2020)
2. Zhang, G.Q., Zhang, Y.S.: Intelligent scheduling algorithm for linear reciprocating track automatic guide vehicle. Comput.Eng. **35**(15), 176–178 (2009)
3. Shen, Y., Guo, B.: The network control system intelligent dynamic scheduling strategy with variable sampling period. J. Sichuan Univ. (Engineering Science edition), **42**(1) (2020)
4. Zha, Z.Y., Li, J.X., Sheng, R.T., et al.: Application of intelligent translational track guide vehicle. Robot. Appl.**5**, 42–43 (2017)
5. Wu, G.H., Zhao, H.N., Zhu, X.C., Wu, Q.: 2019 6th International Conference on Machinery, Mechanics, Materials, and Computer Engineering (MMMCE 2019)

# Chapter 10
# An Apple Grading Method Based on Improved VGG16 Network

**Lin Kaiyan, Zhang Haoping, Liu Chang, Si Huiping, Wu Junhui, and Chen Jie**

**Abstract** Apple grading is an important link of improving the commercial value of apples. This paper proposed an improved convolutional neural network M-VGG16 based on very deep convolutional networks (VGG16) to improve the accuracy and efficiency of apple grading. A convolutional layer of the Conv3, Conv4 and Conv5 was, respectively, deleted compared with VGG16, then a 1 * 1 convolution was connected in parallel with the three groups of convolutions. All convolutional layers were activated by ReLU function. A batch normalization layer was added between each convolutional and activation function layer. Under same dataset, the M-VGG16 network classification prediction accuracy rate is 96.3%, which is higher than AlexNet and VGG16. Compared with traditional machine learning methods, the classification effect of deep learning method has more significant advantages. Therefore, the apple grading method based on deep learning has the advantages of higher grading efficiency and stronger generalization ability, and it provides a new solution for apple classification.

## 10.1 Introduction

Fruit is a major agricultural product in the world. As the largest fruit production and processing country in the world, China is known as the *kingdom of fruit*. Fruit processing has developed to be the third pillar agricultural industry in China with obvious advantages and competitiveness in agricultural product processing industry [1]. Improving the quality of fruits is a key factor for enhancing the competitiveness in the international market. Drawing on the experience of developed countries, fruit grading is one of important ways to increase production value.

Apple classification and rating were usually handled with visual screening which was laborious and time consuming. With the development of machine learning technology, deep learning methods taking advantage of fruit image datasets to mine deep features and had brought new research ideas for solving traditional fruit grading

L. Kaiyan · Z. Haoping · L. Chang · S. Huiping (✉) · W. Junhui · C. Jie
Modern Agricultural Science and Engineering Institute, Tongji University, Shanghai, China
e-mail: sihuiping@tongji.edu.cn

 
L. C. Jain et al. (eds.), *Smart Communications, Intelligent Algorithms and Interactive Methods*, Smart Innovation, Systems and Technologies 257,
https://doi.org/10.1007/978-981-16-5164-9_10

problems [2–5]. Great progress had been made in the field of deep learning methods applied in fruits grading. Sa et al. [6] proposed a method for fruit detection utilizing deep convolutional neural networks and developed an object detector called faster region-based CNN (faster R-CNN) to detect and classify seven types of fruits. This method not only improves accuracy but also allows for rapid deployment. Muresan et al. [7] proposed a deep neural network that was capable of identifying 60 types of fruits. And the paper introduced a high-quality dataset named Fruit-360 that contained popular fruits for obtaining a good classifier. The network finally reached an accuracy of 96.3% on the test set. Kestur et al. [8] presented a *MangoNet* based on architecture of semantic segmentation. Compared with variant architectures of fully convolutional networks (FCN), the *MangoNet* had better detection robustness because it could eliminate the impact of illumination conditions, occlusion in the detection process. The network met the requirements for detection and counting in mangoes yield estimation. In conclusion, the reason why the deep network is faster and more accurate than traditional classification is to improve the architecture in new application scenarios.

With convolutional networks becoming more popular in the field of computer vision, a number of works have been carried out to improve the original architecture. In this paper, we developed a supervised algorithm based on VGG16 in the bid to adapt to the particular fruit classification. The algorithm named modified VGG16 (M-VGG16) could solve the problem of overfitting caused by less data or complicated model in the area of deep learning. Experimental results showed that the M-VGG16 need less time than Alexnet and VGG16 in the training period, but achieving a much higher prediction accuracy.

## 10.2 Material Preparation

### *10.2.1 Apple Grading Standards*

The apple grading standards showed in Table 10.1 were provided by the *National Standard of the People's Republic of China for Fresh Apples* (GB/T10651-2008) [9]. The defects of apple mainly include stab, crush and insect fruit. According to the main features, apples are divided into such four grades as superior grade, first grade,

**Table 10.1** The standards of apple grading

| Features | 1st grade | 2nd grade | 3rd grade | 4th grade |
|---|---|---|---|---|
| Fruit diameter (mm) | ≥100 | ≥85 | ≥70 | <70 |
| Fruit shape | Regular | Slight defect | Defective | Deformity |
| Red color degree (%) | ≥90 | ≥80 | ≥55 | <55 |
| Defective area ($cm^2$) | 0 | 0 | 0 | ≥0 |

(a)1st grade. (b) 2nd grade. (c) 3rd grade. (d) 4th grade.

**Fig. 10.1** Apple images of four grades

second grade and malformed fruit, which are abbreviated as 1st grade, 2nd grade, 3rd grade and 4th grade, respectively.

According to Table 10.1, instead of a single feature, a multi-feature fusion method was beneficial to improve the grading accuracy. Firstly, if one apple has defective area, it will be classified in the 4th grade. If one apple has none defective area, it will be classified according to other characteristics. Secondly, if the other three characteristics are judged to be same rank, the fruit will be classified in this rank. If the above characteristic judgment results are inconsistent, the grade corresponding to the worst characteristic parameter shall prevail.

### *10.2.2 Data Acquisition*

The datasets used in this paper came from two sources. One was to take photos of apples purchased from the fruit wholesale market. Another one came from Web search engine and open-source database. For the first one, images were acquired using the RGB camera. For the second source, images with complete outlines and clearly visible features were selected as part of the dataset.

A total of 3000 apples pictures were collected with each grade having 750 pictures so as to maintain the balance of samples in each category. The image set used for training consisted of 2100 images. The test set was made of 900 images. Examples of apples of different grades are shown in Fig. 10.1.

### *10.2.3 VGGNet*

The VGGNet had confirmed depth was beneficial for the classification accuracy in visual tasks. It utilizes 3 * 3 convolutional layers which are the smallest receptive field. In spite of a large depth, the number of weights is decreased significantly due to a stack of small convolutional filters [10]. Two excellent-performance models (VGG16 and VGG19) were released for accelerating further research. In this paper, VGG16 was chosen as baseline because VGG19 was better suited to complex tasks.

The network structure of VGG16 is shown in Fig. 10.2. The input of the convolutional network is fixed size 224 * 224 * 3 RGB images. Compared with previous work, rather than relatively large receptive fields, 3 * 3 convolution kernels are used

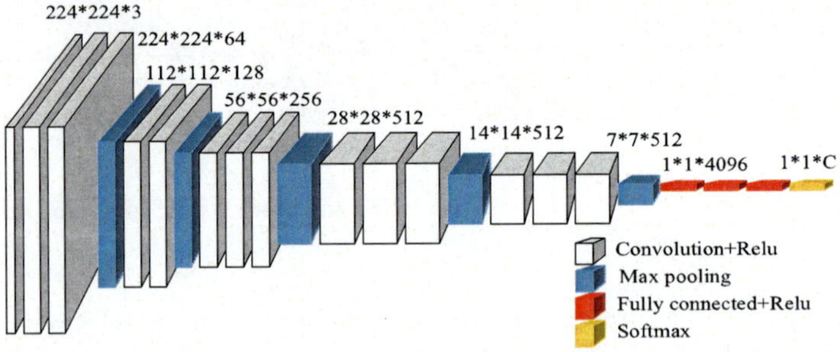


**Fig. 10.2** The network architectures for VGG16

in the whole net. Therefore, the VGG16 increases the number of convolutional layers, making the convolutional neural network deeper and the prediction result more accurate. Furthermore, VGG16 has good generalization ability and show excellent effect when applied to various data types [11].

During training of convolutional neural network, small datasets of labeled images and too complicated models are easier to cause overfitting [12]. Therefore, it was essential for preventing overfitting to collect larger datasets or introduce better techniques.

In our experiment, the size of datasets is small and the resolution of apple images is relatively low. If the network has considerably deep layers, feature information would be lost seriously, resulting in disappearance of the gradient and unsuccess in updating weights of the back propagation. Despite deep convolutional neural networks having attractive qualities, it is obvious to be expensive for a relatively simple task requiring large memory and long training time. The depth VGG16 is moderate, and it is suitable for tasks with fewer samples. In the paper, the VGG16 was selected as the basic network model.

## 10.3 Improvement of VGG16 Architecture

As we have known, it is difficult to train deeper neural networks. A residual learning framework is easier to optimize the degradation problem. With *shortcut connections* to skip one or more layers, a deep residual network could address the notorious problem of vanishing/exploding gradients. The number of parameters is directly determined by convolution kernels size in a convolutional neural network. A big kernel has big perceptual field for recognizing large objections. However, the disadvantage is that the parameters would be redundant [13–15]. Therefore, 1 * 1 convolution filters are often used in terms of network structure design. It performs a linear combination of pixels on different channels, with result in increasing the linear expression and changing the dimension of the feature map. The feature map will be reduced by reducing the dimensionality of 1*1 kernel, and the parameter amount will be reduced accordingly.

The paper proposed a new network called M-VGG16 based on the VGG16 network, the shortcut connections unit is shown in Fig. 10.3. Although the shortcut connections unit added extra parameters, the entire network still had lower complexity and could be easily trained. Therefore, the 1 * 1 kernel could not only provide the previous layer information to the later layer network, but also reduce the parameters in the network and shorten the training time. The overall framework for the M-VGG16 is depicted in Fig. 10.4.

A convolutional layer of the Conv3, Conv4 and Conv5 is, respectively, deleted compared with VGG16, and then a 1 * 1 convolution is connected in the three groups of convolutions. The channels of the three 1 * 1 convolution kernels are 128, 256 and 512, respectively. The output feature maps of the maxpooling layer and that of the 1 * 1 convolution are stacked in series with "concatenate" way so that subsequent network layers could obtain the information of all the previous layers and avoid gradient disappearance. All convolutional layers are activated by ReLU function. Batch normalization (BN) layer is placed between each convolutional layer and activation function layer. The BN changes the distribution of the input value of each layer to standard normal distribution with standardization method so that making the input value fall in the area where the nonlinear function is more sensitive for the input value.

The feature information of different grades of apples is highly similar, so the above improvement method could reduce the loss of feature information and increase the accuracy of model recognition.

**Fig. 10.3** The shortcut connections unit

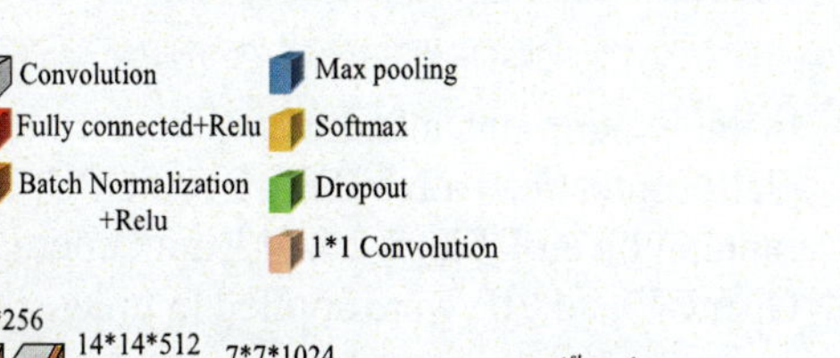


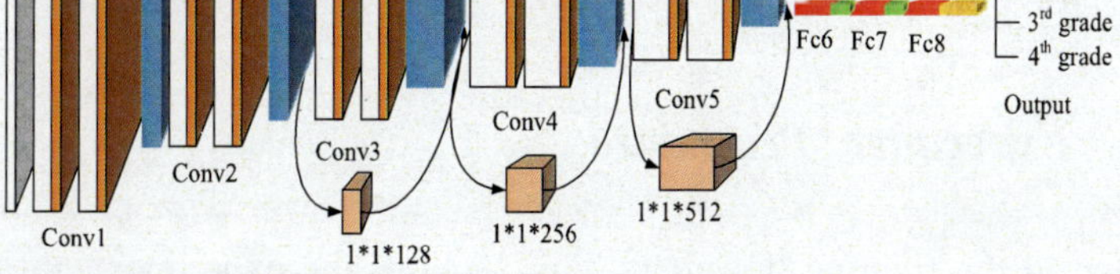


**Fig. 10.4** The network architectures for M-VGG16

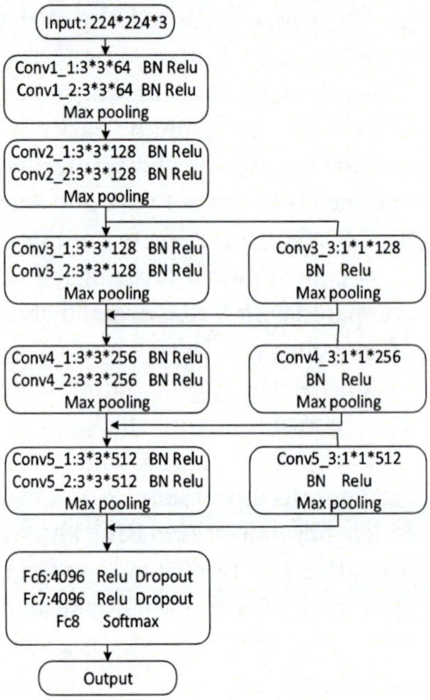


**Fig. 10.5** M-VGG16 model parameter diagram

The parameters of each layer of the M-VGG16 network model are shown in Fig. 10.5 in detail. The initial input image size is 224 * 224 * 3. The final output is a four-dimensional vector, in which each value in represents the probability of one picture belonging to responding category.

## 10.4 Experiment of Apple Grading

### *10.4.1 Hardware Platform*

In this experiment, all networks were trained and tested on a deep learning experiment platform that had a NVIDIA P100 GPU, Intel Core i7 processor and 60 GB memory running on an Ubuntu 16.04 Linux operating system. Open-source libraries such as OpenCV and PIL were applied in image preprocessing. TensorFlow is used to create network model.

### *10.4.2 Experiment Procedure*

As shown in the training flowchart (Fig. 10.6) of the experiment, the process of forward propagation was to extract features, and they were mapped to classification

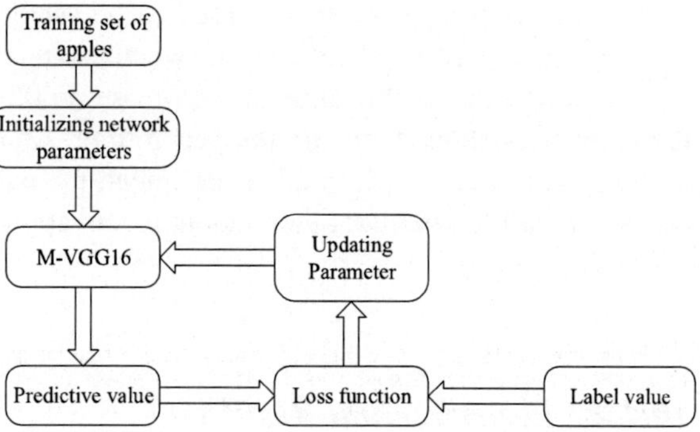


**Fig. 10.6** M-VGG16 training flowchart

results by full connection layer. In the process of back propagation, the errors or loss between the predicted value and the label value was calculated. Here are the specific training steps.

1. Setting input layer size, the kernel size, number and step size of each network layer.
2. Initializing the network parameters according to the standard Gaussian distribution. The mean and variance of parameter distribution are 0 and 1, respectively.
3. Performing the network training in forward propagation. The input images pass through each layer of the network. The prediction result is outputted finally.
4. Calculating the error between the predicted value and the label value and updating the parameters with the gradient descent algorithm.
5. Repeating the above operation to stop when reaching the set training times and saving the model parameters.

### *10.4.3 Training Parameters*

The paper utilized Adam (adaptive moment estimation) optimization method for training. It was an improved stochastic gradient descent algorithm with the advantage that the learning rate could be adjusted for parameters of different frequencies during network training. Weighted sum of cross entropy error function and complex loss were calculated as loss function of the network in this paper.

$$\text{Loss} = -\frac{1}{n}\sum_{p=1}^{n}\sum_{i=1}^{n}[y_i * \log(y_p) + (1 - y_i) * \log(1 - y_p)] + \frac{\lambda}{2}\sum_{i=1}^{n}(y_i - y_p)^2 \tag{10.1}$$

where $n$ is the total number of input samples, $y_i$ is the label corresponding to each sample, $y_p$ is the probability of predicting the input sample as $y_i$, $\lambda$ is the weight of the complexity loss.

Network hyper parameters needed to be set before training the network. The learning rate has a crucial influence on whether a model could be successfully trained. The learning rate for this experiment was set to 0.0001, and batch size represented the number of samples for training each time. An epoch meant that all data in the training set is calculated. In this experiment, the batch size was set to 64 and epoch was set to 200. In addition, the random inactivation probability of the dropout layer was 0.5.

### *10.4.4 Experimental Results*

In this experiment, we utilized the accuracy curve and the loss function curve with corresponding actual score as evaluation index for apple classification. AlexNet and VGG16 were selected to compare with the proposed algorithm. Figure 10.7 depicts

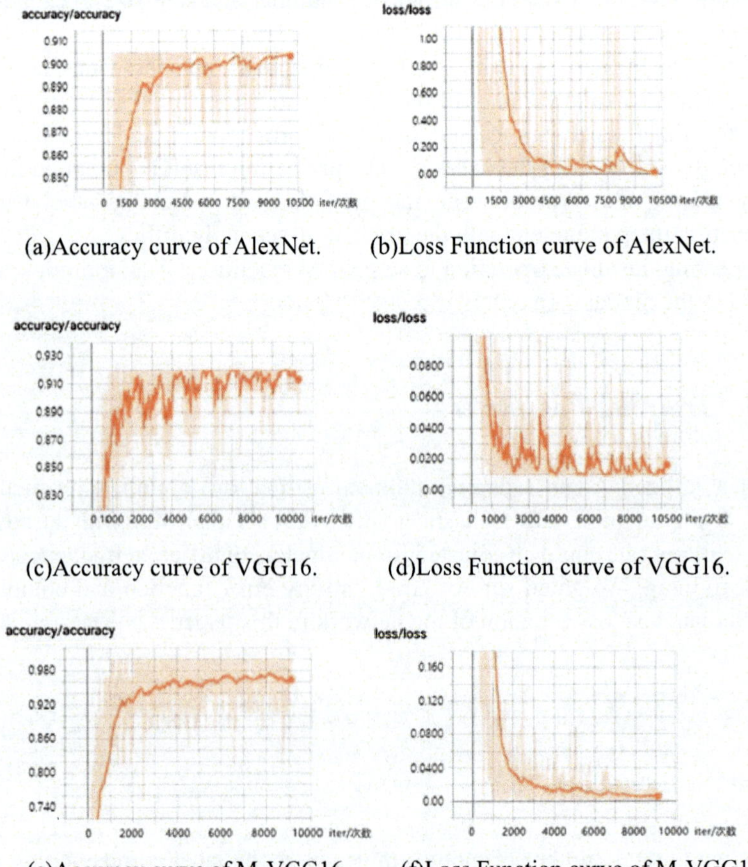


(a)Accuracy curve of AlexNet. (b)Loss Function curve of AlexNet.

(c)Accuracy curve of VGG16. (d)Loss Function curve of VGG16.

(e)Accuracy curve of M-VGG16. (f)Loss Function curve of M-VGG16.

**Fig. 10.7** Accuracy and loss function curve of AlexNet, VGG16 and M-VGG16

**Table 10.2** Grading results of apples

| | AlexNet | VGG16 | M-VGG16 |
|---|---|---|---|
| Training accuracy (%) | 90.5 | 92.4 | 97.8 |
| Test accuracy (%) | 89.7 | 88.9 | 96.3 |

the accuracy and loss function curve of AlexNet, VGG16 and M-VGG16. Train accuracy in the last point and test accuracy are shown in Table 10.2.

As shown in Fig. 10.7a, c, the training time of VGG16 was longer than that of AlexNet. The main reason is that VGG16 has more convolutional layers than AlexNet and the network depth is deeper; therefore, there are more parameters needed to be learned. Moreover, the training accuracy and test accuracy of VGG16 are 0.924 and 0.889, respectively, in Table 10.2. The accuracy of VGG16 is higher than AlexNet in the training set, but lower in the test set.

M-VGG16 obviously outperformed both AlexNet and VGG16 shown in Table 10.2. The proposed method showed that the training accuracy reached up to 0.978. In test set, it could achieve 0.963 of the maximum accuracy and required less computation time. Although M-VGG16 network has the same number of layers as VGG16, it uses 1*1 convolution layer instead of the 3 * 3 convolution layer, which can not only provide the characteristics of the former network layer to the latter for learning, but also make training time shorter. AlexNet and VGG16 had not achieved good results in the task of apple classification. The former had low classification accuracy due to its simple network structure, while the latter caused overfitting problems due to its complex network structure. It is shown that the proposed method M-VGG16 has moderate complexity and significant generalization.

### 10.4.5 Apple Grading Based on Machine Learning Algorithms

In the task of apple grading, color is one of the important criteria for distinguishing apples of different grades. The colors of different grade apple vary greatly. It is appropriate to choose the statistical model of color histogram to grade apples. HSV is the most commonly used color model in image processing, which conforms to human perception of color features. Meanwhile, the shape of an apple is also an important criterion for grade judging. The histogram of oriented gradient (HOG) feature can characterize the shape of an object with the help of the gradient or the directional density distribution of the edge. Support vector machine (SVM) is a supervised machine learning algorithm that realizes feature classification by mapping low-dimensional data to high-dimensional feature space and has strong generalization capabilities. Therefore, this paper extracted the HSV color histogram feature and the direction gradient histogram feature separately as the input of the SVM classifier to distinguish apples of different grades. Firstly, the apple images were converted to HSV space. Secondly, quantizing the three HSV channels and extracting the HSV

**Table 10.3** Results of apple grading experiments

| Method | Index | 1st grade | 2nd grade | 3rd grade | 4th grade | Total |
|---|---|---|---|---|---|---|
| HSV + SVM | NS | 375 | 375 | 375 | 375 | 1500 |
| | FD | 39 | 98 | 18 | 34 | 189 |
| | Accuracy (%) | 89.6 | 73.87 | 95.2 | 90.93 | 87.4 |
| HOG + SVM | NS | 375 | 375 | 375 | 375 | 1500 |
| | FD | 40 | 89 | 10 | 15 | 154 |
| | Accuracy (%) | 89.3 | 76.27 | 97.3 | 96.0 | 89.73 |
| M-VGG16 | NS | 375 | 375 | 375 | 375 | 1500 |
| | FD | 19 | 27 | 9 | 10 | 65 |
| | Accuracy (%) | 94.9 | 92.8 | 97.6 | 97.3 | 95.7 |

The all number of samples (NS), the number of false detection (FD)

color histogram feature vector. Finally, the feature vectors are inputted into SVM for training. On the base of the shape feature classification part, the HOG features of each grade of apples were extracted and input into the SVM classifier for training.

The results are shown in Table 10.3. The overall accuracy based on HSV + SVM and HOG + SVM are 0.874 and 0.8973, respectively. Two machine learning algorithms have achieved relatively good classification results, while the classification accuracy rate based on the convolutional neural network M-VGG16 reaches 0.957. Compared with these two traditional machine learning methods, the classification effect of deep learning methods has more significant advantages.

## 10.5 Conclusions

This paper studied the apple grading method based on deep learning. In order to solve the problem of overfitting and loss of feature information, this paper proposed an improved neural network M-VGG16. VGG16 was selected as the basic framework to reduce the complexity of the network. 1 * 1 convolution layer was connected in parallel next to the Conv3, Conv4 and Conv5 in the backbone network to reduce the loss of feature information. The three network architectures were tested on the same apple dataset. Comparing with AlexNet and VGG16, the M-VGG16 had the best classification effect and the fastest network convergence rate. It effectively solved the problem of overfitting. Moreover, this article compared the performance of deep learning and traditional classification methods. The conclusion was that the M-VGG16 method based on deep learning has the advantages of higher grading efficiency, stronger generalization ability and automatic feature extraction in apple grading.

**Acknowledgements** This work was supported by a National Science and Technology Support Project named "Research on Evaluation Index and Standard System of Green Technology" (Project No. 2017YFC0212901).

## References

1. Sui, W., Jia, H.: Research on utilization status and classification management of solid waste in fruit processing in China. Trans. Chin. Soc. Agric. Eng. **34**(S1), 172–180 (2018)
2. Huang, C., Fei, J.: Online apple grading based on decision fusion of image feature. Trans. Chin. Soc. Agric. Eng. **33**(1), 285–291 (2017)
3. Gu, Y., Shi, G., Liu, X.: Optimization spectral clustering algorithm of apple image segmentation with noise based on space feature. Trans. Chin. Soc. Agric. Eng. **32**(16), 159–167 (2018)
4. Kumar, A., Kumar, K.: Time-frequency analysis and support vector machine in auto-matic detection of defect from vibration signal of centrifugal pump. Measurement **108**, 119–133 (2017)
5. Jaffar, A., Jaffar, R., Jamil, N.: Photogrammetric grading of oil palm fresh fruit bunch-es. Int. J. Mech. Mechatron. Eng. **9**(10), 18–24 (2017)
6. Inkyu, S., Zongyuan, G., Feras, D.: DeepFruits: a fruit detection system using deep neu-ral networks. Sensors **16**(8), 1222–1228 (2017)
7. Mureşan, H., Oltean. M.: Fruit recognition from images using deep learning. Acta Uni-versitatis Sapientiae Informatica **10**(1), 26–42 (2018)
8. Kestur, R., Avadesh, M., Omkar, N.: Mango Net: A deep semantic segmentation archi-tecture for a method to detect and count mangoes in an open orchard. Eng. Appl. Artif. Intell. **77**, 59–69 (2019)
9. State Administration for Market Regulation, Standardization Administration.: GB/T1065-2008, Fresh apple. China Federation of Supply and Marketing Coopera-Tives (2008)
10. Guo, Z., Chen, Q., Zhang, B.: Design and experiment of handheld near-infrared spec-trometer for determination of fruit and vegetable quality. Trans. Chin. Soc. Agric. Eng. **33**(8), 245–250 (2017)
11. Sapan, N., Bankim, P.: Usage of image processing and machine learning techniques in agriculture—fruit sorting. CSI Commun. **37**(7), 25–27 (2013)
12. Blasco, J., Aleixos, N., Moltó, E.: Computer vision detection of peel defects in Citrus by means of a region oriented segmentation algorithm. J. Food Eng. **81**(3), 535–543 (2007)
13. Xianfeng, L., Weixing, Z.: Apple grading method based on features fusion of size, shape and color. Procedia Eng. **15**, 2885–2891 (2011)
14. Wang, Z., Chi, Q., Guo, W.: Internal quality detection of apples during late development period based on near-infrared spectral technology. Trans. Chin. Soc. Agric. Mach. **49**(05), 348–354 (2018)
15. Devrim, U., Bemard, G., Olivier, K.: Automatic grading of Bi-colored apples by multi-spectral machine vision. Comput. Electron. Agric. **75**(1), 204–212 (2011)

# Chapter 11
# Comprehensive Application of Big Data in Lean Management of Distribution Network Operation

**Jieming Zhang, Weiwen Li, Yuqian Chu, and Yanzhi Liang**

**Abstract** In order to solve the problems of distribution network, equipment, management difficulty, low-real-time distribution network information, this paper systematically analyzes the current situation of distribution network operation and maintenance. As well as the application and research of data mining and analysis processing technology in distribution network, big data mining application is introduced into daily operation and maintenance of distribution network, and lean management is carried out to strengthen the effective control and control of power network from the source, so as to eliminate waste and create value according to 'eliminate waste, create value, continuous improvement, excellence' requirements, all.

## 11.1 Introduction

In view of the complex data source, large amount of data and difficult data integration in distribution network, the application analysis and research of distribution network big data will carry out with "multisource data integration, multipath data audit, multidisciplinary information fusion" as the core to build the lean management system of distribution network, which provides scientific guidance and effective basis for distribution network planning, customer access planning and management. At present, the main problems of regional distribution network are as follows.

J. Zhang (✉) · Y. Chu · Y. Liang
Zhaoqing Power Supply Bureau of Guangdong Power Grid Co., Ltd., Zhaoqing 526060, Guangdong, China
e-mail: zqjamin@163.com

W. Li
Zhaoqing Deqing Power Supply Bureau of Guangdong Power Grid Co., Ltd., Zhaoqing, 526600, Guangdong, China

 
L. C. Jain et al. (eds.), *Smart Communications, Intelligent Algorithms and Interactive Methods*, Smart Innovation, Systems and Technologies 257,
https://doi.org/10.1007/978-981-16-5164-9_11

### *11.1.1 Topological Relationship Verification of Distribution Network Station Area is Time-Consuming and Laborious*

In order to reduce the wattful power network loss, the network structure of the distribution network needs to be adjusted [1]. However, after the adjustment of the network structure, the relevant information in the electricity consumption information collection system is usually checked and updated by manual means, which is time-consuming and laborious. The correctness of distribution network line variation relationship has a great significance to the network loss calculation and the distribution network faults repair. It is necessary to ensure the correctness of distribution network line variation relationship through verification.

### *11.1.2 The Operation and Maintenance Management of Existing Distribution Station Area not Enough Lean*

(1) The lack of important parameter information such as connection groups and the number of taps in distribution station area is serious. Yyn0 distribution station area three-phase imbalance causes export problem [2]. According to statistics, due to the imbalance of three-phase load in Yyn0 distribution station area, the adjustment of three-phase load imbalance does not reflect the difference; it is difficult to obtain the position of transformer tap in distribution station area, about 70% of the low-voltage station area is in rated transformer tap, therefore, the transformer can be adjusted in low-voltage station area. The partial data of distribution station area capacity is not true and accurate, so it is difficult to monitor the load of distribution station area effectively.

(2) Unable to effectively know the operation status of the distribution station area. The number of distribution station area is huge under the limited human resources. It is difficult to locate the specific station area, such as overheating of joints, abnormal voltage classification, high-risk failure during overload, lead to distribution network operation and maintenance of blindness [3].

(3) Do not consider the optimal strategy of distribution station area load adjustment, transformer tap adjustment and others. Under the limited human resources, and the number of distribution transformers is huge, the differential operation and maintenance management can't be realized, which may cause the distribution station area to adjust transformer tap repeatedly, maintenance staff has a heavy workload.

### 11.1.3 The Technical Reconstruction Projects in the Existing Distribution Station Area Are not Accurate Enough

Due to the lack of data support and systematic evaluation system and method, using extensive and qualitative evaluation methods to evaluate:

(1) In the aspect of investment decision, it mainly focuses on the demonstration of the necessity and feasibility of the project, and lack of specific demonstration on the investment benefit of the project, only according to the historical data to carry on the simple proportional distribution [4].

(2) In the aspect of technical reconstruction effect evaluation, it mainly carries on the statistics of scale increase. The data lack of objectivity and accuracy, because the data source reported from grass-roots team. The technical reconstruction has the characteristics of small scale, numerous and complicated; therefore, it is difficult to invest and evaluate effect.

With the improvement of lean management in distribution network, it is very urgent and important to carry out accurate investment and effectiveness evaluation in distribution station area.

### 11.1.4 The Requirement of IT Capability for Mass Data Query

See Table 11.1.

**Table 11.1** Existing problems and effects

| Existing problems | Effects |
|---|---|
| Low-data diversity: not include diversified data such as SCADA system, operation management system (OMS), production management system (PMS), geographical information system (GIS), marketing management system (including 95,598), electricity information acquisition system, power distribution production and emergency repair platform, etc. | The single-dimensional statistical analysis cannot fully reflect the real state of the distribution network |
| Lack of data accuracy | Insufficient validity of data analysis |
| The rapid increase in data capacity leads to the slow operation of the existing data analysis model | Low-data utilization |

## 11.2 Data Mining and Research of Distribution Network

### *11.2.1 Deep Mining the Big Data of Operation and Power Distribution System to Improve the Load Forecasting Ability of Different Industries and Regions*

Based on the statistical analysis of the big data of operation and power distribution system in the database of multi-source distribution network, the power consumption characteristics of different industries and regions are analyzed. Considering the influence of external economic environment on electric quantity, focuses on the power load predictive ability and management capability to establish an optimization model.

The load history curve of users is analyzed, for example, calculating the maximum load and minimum load (day, month, season, year), load rate and imbalance coefficient, an automatic identification method is proposed which can automatically identify the stage characteristics of load, the capacity of power consumption, the location information and so on. This paper analyzes the data of distribution network statistical database (including power information acquisition system, load control system, distribution transformer online test system, etc.), and extracts the basic database of power consumption characteristics [5, 6].

Using time series analysis of user power consumption data to summarize the load characteristics of different industries in different time dimensions, combined with the user's typical service factor and phase characteristics to form typical load characteristic curve of different industries. Various types of user load curves are superposed to form typical load characteristic curve, and a composite model is established.

### *11.2.2 Deep Mining the Load Data to Improve Access Ability of Distribution Network*

Lean research on the access capacity of distribution network planning, put forward the load demand coefficient and line assembly capacity standard, and models and realizes the theoretical analysis, which provides scientific guidance for distribution network line planning and user access planning [7]. In terms of various load characteristics analysis, it is divided into two levels: distribution line and transformer. Based on the statistical analysis of annual load characteristic curve and the daily load characteristic curve, the load classification and load characteristics of distribution users are summarized. At the same time of big data mining, taking others as reference, the development trend of load demand coefficient of typical distribution network users is investigated, the correction conclusion is further verified, and the recommended value of load demand coefficient is determined [8].

## 11.3 Lean Management Platform of Distribution Network

### *11.3.1 Three-Phase Imbalance Analysis of Distribution Network*

Find out the reason of three-phase imbalance and the long-term imbalanced distribution transformer to realize the comprehensive and real analysis of the distribution transformer three-phase imbalance. We excavate the user's electricity consumption behavior, analyze the change trend of user's electricity consumption load and distribution load, and construct the influence of user's electricity consumption behavior on three-phase imbalance.

### *11.3.2 Voltage Quality Analysis of Distribution Network*

Realize multi-dimensional statistical analysis of voltage quality (according to the number of out of limit, duration, reasons, etc.), analysis of distribution transformer connection group and in operation gear. Mining the reason of voltage out of limit, record and summarize the distribution transformer with unreasonable reactive power compensation and operation gear [9]. Establish long term low-voltage user correspondence table of distribution transformer, provide low-voltage reconstruction scheme, and adjustment scheme of voltage optimization (such as gear adjustment). Control the gear adjustment during peak demand for electricity to improve voltage qualification rate [10].

### *11.3.3 Load State Analysis of Distribution Network*

Based on the equipment record and telemetry data, the real and comprehensive control of load state (heavy load, overload, light load, no load, load rate, etc.) at each level (distribution feeder/branch line/low voltage line/distribution transformer) in the distribution network, and the multi-dimensional statistical analysis can be realized (such as by time, reason, load trend) [11]. We dig for the cause of overload, terminal abnormality, equipment record, etc. Reasonably arrange the operation and maintenance plan of distribution transformer to reduce heavy load and overload, realizing economic operation.

### 11.3.4 Low-Power Factor Analysis of Distribution Network

Based on the equipment record and telemetry data, the multi-dimensional statistical analysis of power factor, reactive power and active power of distribution transformer (such as by time period, region and distribution transformer type) are realized, and the power factor of customers (station area) is truly and comprehensively controlled, And improvement the potential of customers (station area) power factor automatically, assist and guide the high loss customers (station area) to carry out energy-saving optimization.

### 11.3.5 Reactive Power Operation Evaluation of Distribution Network

Based on the network topology relationship and operating data, evaluate the voltage/reactive power control level, and realize the multi-dimensional statistics of power factor qualification rate, reactive power control over/under compensation rate, power factor load matching degree and capacitor/reactor action times exceeding the standard [12]. The supply and demand relationship of reactive load at all levels are controlled truly and comprehensively, realize the evaluation of reactive compensation capacity configuration at each level, multi-dimensional statistics of total capacitive/inductive reactive power capacity compensation degree, load matching degree, capacity shortage and the improper rate of a single-group capacity configuration of capacitive/inductive reactive power compensation equipment, to assist each unit in reactive power planning and optimization and improve the economic operation level of the distribution network.

## 11.4 State Assessment of Distribution Network Equipment

### 11.4.1 State Assessment of Distribution Area—Contact State Assessment

The technical scheme of data-driven distributed area contact state assessment is as follows. First of all, it selects 10 kV line voltage curve as the standard curve. Secondly it considers the three-phase voltage of each collection point is basically unequal under the condition of three-phase imbalance of Yyn0 distribution area. The reduction is performed to the outlet voltage of Dyn11 and Yyn0 distribution area [13], and it finally carries out the correlation calculation to the voltage curve and 10 kV line voltage curve of tested distribution area. The larger the absolute value of correlation coefficient, the stronger the correlation for the tested distribution area and 10 kV line voltage curve, as well as the smaller the possibility for poor contact of tested distribution area. The closer to 0 the absolute value of correlation coefficient, the weaker the correlation between tested distribution area and 10 kV line voltage curve, as well as the larger the possibility for poor contact of tested distribution area [14].

Data-driven contact state assessment technology of distribution network is widely used. It utilizes the massive time series data of data service center area to carry out the on-line assessment to the contact state of running distribution transformer. It is able to proactively identify the poor contact of distribution area, realize the secondary development and application for time series data of area, and effectively enhance the safe operation level of low-voltage distribution network.

### *11.4.2 Operational Status Assessment of Area-Overload Fault Risk Assessment*

Impact of distribution transformer overload on safe operation of area. Under the different overload coefficients and environmental temperature, whether the distribution transformer of area allows the overload will not impact the service life of distribution transformer or cause the overload damage of distribution transformer.

Risk assessment technology of distribution transformer overload fault. The distribution transformer can operate for a long time under the normal working condition, but when the distribution transformer is overload operating for a long time, the temperature of winding insulation will be increased. When the winding hot-spot temperature exceeds 140 °C, bubbles will be generated, which will reduce the insulation strength of distribution transformer. It is possible to reach the degree of damaging the winding insulation. When the distribution transformer is overload operating for a short time, the insulation strength may appear the temporary reduction. When the overload is reduced to the normal level, the service life of distribution transformer basically will not be impacted. Therefore, the overloading capacity of distribution transformer needs to be determined by its top oil temperature and winding hot-spot temperature [15].

## 11.5 Conclusion

It is required to establish the overall-process guidance and assessment system including information management of distribution area-lean operation and maintenance of distribution area-precise technical modification of distribution area. Based on the operation date of accessed distribution network and equipment, according to the specific application scene of distribution area, it designs the reasonable data-driven model. And then according to the data-driven results, it realizes the lean operation and maintenance and precise modification of distribution area, as well as performs the scientific and quantitative assessment to the operation, maintenance and technical modification results of distribution area. The former qualitative evaluation method is changed, and the scientificity, rationality and economy of distribution network operation, maintenance, technical guidance and evaluation are improved. The data-driven system, result assessment model, assessment method and auxiliary decision-making system raised by the project have a strong operability, which can be as the evidence for the guidance and assessment of operation, maintenance and technical modification for the distribution area of all provincial power network companies, prefecture companies and country-level power supplying enterprises.

### *11.5.1 Enhance "Lean Management" for Operation and Maintenance of Distribution Area*

Through the assessment and analysis to the application data-driven model of current status for operation and maintenance of distribution area, it is able to grasp the current status for operation and maintenance of distribution area. According to the specific application scene of distribution area, it is able to provide the optimal differential operation and maintenance auxiliary strategy, change the traditional operation and maintenance method based on experience, greatly improve the pertinence for operation and maintenance of distribution area, effectively reduce the manpower operation and maintenance workload of operation and maintenance team, as well as avoid the blind operation and maintenance, ineffective operation and maintenance of distribution area. Based on the current status for operation and maintenance of distribution area, as well as the corresponding optimal reasonable operation and maintenance strategy, it is able to grasp the assessment of actual results for operation and maintenance of distribution area so as to enhance the lean operation and maintenance level of distribution area.

### *11.5.2 Realize Modified "Precise Investment" of Distribution Area*

Based on the distribution network, fundamental record and operational data of equipment, it is able to identify the risk hazard and weak equipment, e.g., long-term overload, low voltage, imbalance of three-phase, bottleneck and frequency power outage, it establishes the technical modification program series of distribution area, and provides the modification plan for the equipment having problems, as well as lists modification demand base so as to provide the auxiliary strategy support. It establishes the assessment model of technical modification results for data-driven distribution area performs the technical modification investment analysis of distribution area on a regular basis, as well as streamlines the direction and key point of investment, which can effectively resolve the outstanding problems of vague direction of technical modification investment of distribution area, unreasonable investment structure and low profit of investment, etc.

## References


1. Xu, H., Zhu, T., Chen, X.: Line planning of 10 kV distribution network. Guangdong Electric Power **30**(7), 12–17 (2017)
2. Gao, Q., Xu, C., Chen, X.: Lean management of distribution network based on big data analysis rural electrification 3 (2016)
3. Kong, H.D.: Research and example of lean management in distribution network. Electr. Appl. **11**, 58–62 (2015)
4. Zhang, J., Wang, C.: Analysis on the present situation and development trend of distribution network. Modern Ind. Econ. Inf. **5**, 31–32 (2015)
5. Hu, L., Diao, Y.: Reliability analysis of distribution network operation based on big data technology. Power Netw. Technol. **41**(1), 265–271 (2017)
6. Liu, K.: Research on big data application requirements and scenario analysis of intelligent distribution network. Power Netw. Technol. **35**(2), 287–293 (2015)
7. Song, Y.: Research and application of data integration technology for integrated distribution network planning and design platform. Power Netw. Technol. **40**(7), 2199–2205 (2016)
8. Zhang, Z.: Application exploration of big data in lean management of distribution network. In: 2017 Smart Grid Conference on "Electronic Technology Applications" (2017)
9. Zhang, W., Zhang, N.: Research on lean management of distribution network based on big data operating condition control platform. Autom. Instrum. 7, 98–103 (2018)
10. Zhao, X.L.: Post Evaluation research on Big Data Processing and Scheduling Support Project of Taiyuan Distribution Network. North China Electric Power University (2015)
11. Design and application of intelligent operation and maintenance control platform for distribution network based on big data platform. Electric Autom. **40**(6), 85–88 (2018)
12. Liu, J., Li, Q.: Analysis and application of distribution network lean management mode based on real-time control platform. Electrotechnics Electric. **8**, 65–68 (2017)
13. Hu, J.D., GAO, Y.X.: Research on input-output benefit analysis and decision-making technology of distribution network based on big data. Chinese Sci. Technol. **1**, 87–91 (2017)
14. Research on typical application scene of big data in intelligent distribution network. Power Syst. Big Data **21**(11), 49–55 (2018)
15. Discussion on application function architecture of intelligent operation and maintenance control platform for distribution network. Electromech. Inf. (2017)

# Chapter 12
# Research on the Construction of Intelligent Vehicle Verification System for Road Transportation

**Haiying Xia, Shuquan Xu, Yingji Liu, Xinlei Wei, and Hong Jia**

**Abstract** The technical management of road transport vehicles is an important work to ensure the safety of road transportation. To manage the vehicles entering the road transport market from the source of production, so that the quality and performance of vehicles should meet the requirements of relevant standards, and be able to drive safely on the road is the main purpose of implementing the standard management system for road transport vehicles. It is of great significance for improving the management level and service quality of road transportation industry to establish an intelligent vehicle verification system for road transportation standards, which can effectively improve the efficiency of vehicle verification. Artificial intelligence technology is used to realize intelligent audit, automatic identification and comparison of announcement parameters. In the later stage, data mining technology is used to mine data rules and values, which provides decision support for more scientific and reasonable management of road transport vehicle production and verification, and improvement of relevant laws and regulations.

## 12.1 Introduction

### *12.1.1 Industry Background*

According to the requirements of relevant national laws and regulations, the Ministry of Transport implements the standard management system for road transport vehicles, with the main purpose of improving the technical level of "safety, efficiency and energy saving" of road transport equipment, and ensuring that the quality and performance of vehicles entering the road transport market should meet the relevant standards. In order to realize the scientific and reasonable application of the "Announcement of Road Transport Standard Vehicles" and ensure the closed

H. Xia (✉) · S. Xu · Y. Liu · X. Wei · H. Jia
Key Laboratory of Operation Safety Technology On Transport Vehicles, Ministry of Transport, Beijing 100088, China
e-mail: hy.xia@rioh.cn

 
L. C. Jain et al. (eds.), *Smart Communications, Intelligent Algorithms and Interactive Methods*, Smart Innovation, Systems and Technologies 257,
https://doi.org/10.1007/978-981-16-5164-9_12

management system of precision service industry, it is necessary to build an intelligent vehicle verification system for road transportation. The system is of great significance to improve the verification efficiency of road transport standard vehicles, protect the legitimate rights and interests of consumers, effectively promote the process of road transportation informatization construction, and improve the management efficiency and service quality of road transportation industry. It is an important measure to effectively promote the implementation of road transport standard vehicle management.

### 12.1.2 Situation Analysis

In order to further improve the quality and informatization level of real vehicle verification of road transportation standard vehicles, the Research Institute of Highway Ministry of Transport has developed a unified verification system for road transport standard vehicles on the basis of the original single-announcement query. The system has the functions of vehicle identification, online verification, automatic comparison and announcement, statistical analysis, report management and electronic fence and other functions.

The system includes intelligent audit works. In fact, the audit work is a lot of repetitive work, and these audit indicators have a standard range, there are certain rules, so the use of artificial intelligence technology to achieve intelligent audit can improve audit efficiency. Some researchers have proposed an intelligent model for the intelligent review of bidding documents. This method is based on the method of text mining [1]. At present, the intelligent audit system of medical system is widely and deeply used. In the hospital, there are many items of traditional Chinese medicine detection, including indicators, medical orders of text data type, image data and other cross media data. Using natural language processing and semantic recognition technology can improve the efficiency of intelligent audit. In reference [2], a medical laboratory intelligent audit model based on machine learning was proposed. The model was a combination of extreme gradient boosting and gradient boosting decision tree + logistic regression, the blood routine test was taken as the research object, and the audit accuracy rate reached 99%. Intelligent audit is not only widely used in medical laboratory audit, but also widely used in medical insurance audit [3]. Literature [4] pointed out that the research on the safety technology audit of operating vehicles is not optimistic, and analyzes the safety technology status of operating vehicles and puts forward countermeasures. However, these are only from the perspective of policy to optimize the safety technology inspection of operating vehicles, and do not substantially improve the efficiency of compliance audit of operating vehicles from a technical point of view.

Data mining technology has gradually formed a set of basic theoretical basis, mainly including classification, clustering, pattern mining and rule extraction. By analyzing the types and distribution of abnormal data, and based on the mathematical and logical relationship between business data, reference [5] puts forward the screening rules and processing methods of abnormal data, and gives the system optimization strategy scheme. Reference [6] analyzes the quality data collected by the vehicle off-line detection system. At home and abroad, a large number of research results has been achieved in the theoretical aspect of association rule mining [7]. Association analysis is used to explore the correlation between research objects, which is generally realized by mining frequent item sets in data [8]. Data mining technology is used to deeply analyze and process the uploaded data of vehicle verification system, screen false reported data and non-standard data, mine and summarize the objective laws behind the data, and provide support for the formulation of vehicle standards and specifications.

## 12.2 System Workflow

As shown in Fig. 12.1, in terms of system implementation procedures, the Ministry of Transport is responsible for the announcement management of vehicle models at the source of production. It mainly carries out the third-party performance test, technical review and announcement of sample vehicles according to relevant standards, and uniformly publishes the announcement data of up to standard models through the "Road Transport Vehicle Technical Service Network." This provides a technical basis for the local road transport management departments to carry out the verification of vehicles up to standard and handle the "Road Transport Certificate."

As shown in Fig. 12.2, from the perspective of system implementation, since the implementation of standard vehicle model system, it has indeed played an important role in promoting the technical upgrading of road transportation equipment, but in the specific implementation process, there are still many quality management problems. In the actual verification of qualified vehicle models, local transportation management departments found that the production of some vehicles did not conform to the actual announcement, and the vehicle performance did not meet the standard requirements. These problems are mainly reflected in the following aspects: the external dimensions, the internal dimensions of the freight car or the height of the guardrail, the engine model and other parameters are inconsistent with the announcement of the standard model. That is to say, after passing the announcement of standard models,

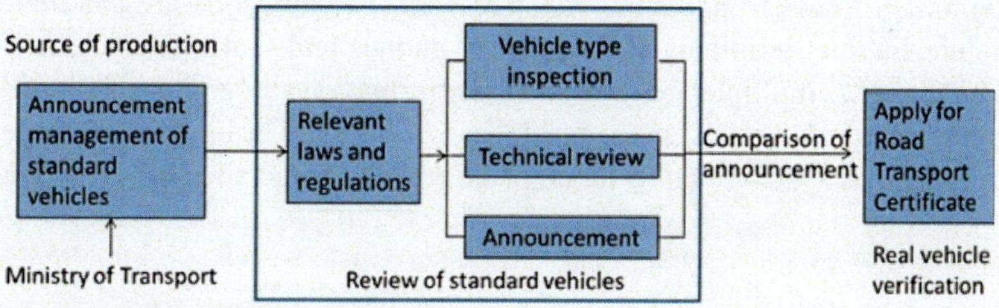


**Fig. 12.1** System implementation procedure

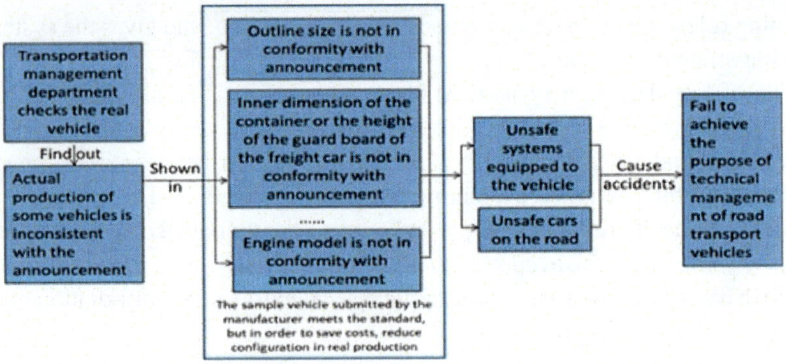


**Fig. 12.2** Verification of implementation

some automobile manufacturers fail to produce according to the standards of sample cars. In order to reduce costs, they reduce the allocation and lower the production standards of vehicles, there may be "unsafe system equipped" or "unsafe vehicles on the road." There are many potential accidents and safety problems. Objectively speaking, the purpose of transport vehicle announcement management will not be achieved.

From the perspective of supervision and management, there is no effective information feedback channel for the management of up to standard vehicle models at present. The problems found in the transportation management verification in various regions cannot be timely and effectively fed back. There is no management chain of announcement information disclosure of standard vehicles, verification of standard vehicles, problem feedback and rectification, which cannot form a good reverse mechanism for automobile enterprises.

## 12.3 System Construction

### *12.3.1 Overall Functional Structure*

As shown in Fig. 12.3, the system includes verification application app and the PC terminal of background supervision system. The app software can be installed in mobile Internet terminals such as smart phones and tablet computers. The app terminal includes multiple modules such as information automatic collection, logical judgment, picture/video forensics, result generation, etc. The background supervision system includes system management, background storage and statistical analysis.

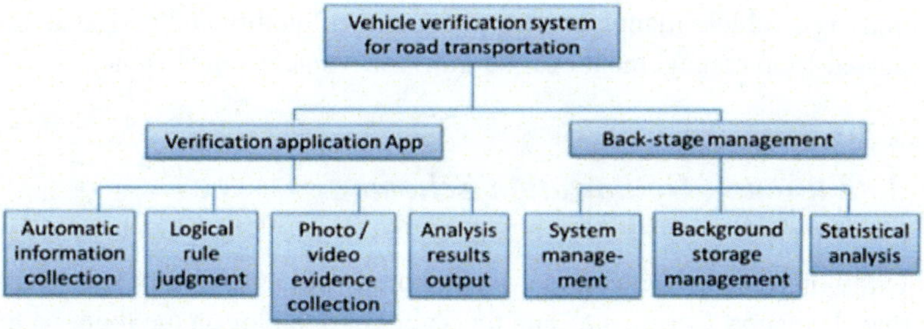


**Fig. 12.3** Overall functional architecture

## *12.3.2 Realization of Specific Functions*

The specific functions of the system are as follows.

(1) System management includes user management, authorization management and other functions. It is mainly used to set and allocate system users' permissions, as well as necessary records and information backup of operation behaviors.
(2) During the on-site verification, when the user enters or automatically identifies the quick response code or vehicle identification number of the vehicle to be verified, it automatically returns the factory configuration parameter data of the vehicle, so as to reduce the information input workload of the verification user and avoid errors. The system is encapsulated with logic judgment rules designed in accordance with the requirements of verification standard. It can automatically judge whether it meets the verification requirements, and push the results back to the verification user. In the process of on-site verification, the verification user can use the system to take photos or videos and upload them as the basis and evidence of verification.
(3) The system can use mobile network or Wi-Fi to upload the whole verification result information in real time. The on-site inspectors and backstage managers can query the verification information. The system generates the verification result data of road transport vehicles, which can be imported into the transportation administration system in various forms.
(4) The system has been applied in Guizhou, Sichuan, Shandong, Zhejiang, Heilongjiang and other provinces, and has achieved good demonstration effect. At present, nearly 150 thousand new car access checks have been completed in the verification system, and the one-time pass rate is about 90%. At the same time, with the continuous accumulation of business data, big data analysis began to play a role. Through the system, we can effectively analyze the common types of unqualified items in the process of passenger car and freight car verification and the proportion in all verified vehicles. Further help the industry management departments to focus on the feedback of these problems,

and urge vehicle manufacturers to strengthen quality control, so as to form a closed-loop management of road transport vehicles up to standard.

### *12.3.3 Function Optimization Scheme*

The quality of vehicle verification system directly affects the efficiency of applying for "Road Transport Certificate" and the safety of vehicles on the road. At the same time, the verification data quality analysis is also the key technical difficulty to be solved. It is very necessary to study the quality monitoring and analysis methods of road transportation vehicles. By means of information technology and big data mining technology, statistical analysis is carried out on the verification data of qualified vehicles to effectively identify and judge the low-quality and false verification records, and to accurately control the inspection quality of road transportation standard vehicles.

(1) Design policy interpretation and laws and regulations query function, can query various standards in time, make it convenient for the vehicle to be checked to make timely judgment according to the standard. The answers to the practical problems are summarized and reflected in the app.
(2) Lock the vehicles that fail to pass the one-time inspection. If the testing organization is replaced, it is necessary to give a warning and prompt to strictly review, and fraud is not allowed.
(3) Strengthen the function of data statistics and analysis, and summarize the classified information of substandard and up to standard vehicles, so as to facilitate the management department to know more about the specific situation of vehicle verification in this region.

## 12.4 Conclusion

To sum up, optimizing the intelligent vehicle verification system for road transportation can better guarantee the quality of vehicle verification data, promote the quality management of automobile production enterprises, and improve the work efficiency of the verification management department. It is a good means in the quality control technology of road transportation vehicles, which can provide better technical service support for all levels of transport management institutions, vehicle inspection and testing institutions and automobile production enterprises.

**Acknowledgements** This paper is supported by the Central Public Research Institutes Special Basic Research Foundation (2020-9065).

## References


1. Yingying, C., Li, Z., Feng, Z.: Intelligent model of bidding audit based on text mining. In: 2018 IEEE International Conference of Safety Produce Informatization (IICSPI). IEEE (2018)
2. Junlong, L., Xiaohong, X.,Huazhen, W.: Intelligent verification model of medial examination based on machine learning. China Digital Med. **14**(03), 53–55 (2019)
3. Sijie, L., Qiaoshu, W., Xueqing, G.: Design and application of the intelligent audit platform based on medical insurance rules. China Digital Med. **12**(7), 69–71 (2017)
4. Xiuming, Z.: Detection and analysis of safety technology status of operating vehicles and Countermeasures. Heilongjiang Sci. Technol. Inf. **32** (2016)
5. Lu, Z., Yanxia, C., Jie, Z.: Data quality analysis and improved strategy research on operations management system for electric vehicles. In: Proceedings of IEEE International Conference on Electric Utility Deregulation, Restructuring Power Technology, DRPT (2015)
6. Huangzheng, G., Yufeng, D., Changbinm G.: Quality analysis and process optimization of vehicle production off-line based on data drive. Technol. Dev. Enterp. **38**(06), 16–20 (2019)
7. Zhenwei, Y.: The Research of Data Mining Algorithm in Logistics Transportation Based on Association Rules. Beijing University of Posts and Telecommunications, Beijing (2017)
8. Ruojia, W., Siyi, W., Yiran, Z., Jimin, W.: Review of data mining techniques application in medical and healthcare field. Intell., Inf. Sharing 05 (2018)

# Chapter 13
# Quantitative Risk Analysis of Dangerous Goods Road Transport Vehicles and Networks

**Xuan Dong, Zhong Xu, and Li Yuan**

**Abstract** The risk analysis of dangerous goods is the main basis for making all decision-making of dangerous goods transportation. The accuracy and practicability of risk analysis model directly determine the level of dangerous goods transportation management. On the basis of summarizing the research results of risk analysis of dangerous goods transportation at home and abroad, a quantitative analysis model of dangerous goods transportation suitable for China's national conditions was proposed. The risks caused by non-traffic accidents and black spots were included in the model for the first time, and TransCAD was used to apply the model to the risk analysis of actual road networks.

## 13.1 Introduction

In recent years, with the rapid development of China's social economy, the promotion role of transportation in social and economic development is more and more obvious, especially the rapid development of highway construction in the guarantee of China's rapid and healthy economic development has played an important role. After more than a decade of development, China had 149,600 km of expressways open to traffic by the end of 2019, basically realizing inter-regional connectivity and forming a road network structure. With the formation of expressway network structure, its operation and management mode is gradually changing from single line relatively independent operation to multi-line comprehensive operation, forming a new situation of expressway network operation. Compared with the previous single operation, the characteristics of highway network operating mainly in the road network scale expands unceasingly, the obvious contradiction between road network traffic capacity and traffic demand, road traffic accidents and severe degree is high, total local

X. Dong (✉)
A Key Laboratory of Operation Safety Technology On Transport Vehicles, Ministry of Transport, PRC No. 8, Xitucheng Road, Haidian District, Beijing 100088, China

Z. Xu · L. Yuan
Shenzhen Yuwei Information and Technology Development Co.,Ltd, Floor 3, Building F XingHe WORLD No.1 YaBao Road LongGang District, Shenzhen, People's Republic of China

 
L. C. Jain et al. (eds.), *Smart Communications, Intelligent Algorithms and Interactive Methods*, Smart Innovation, Systems and Technologies 257,
https://doi.org/10.1007/978-981-16-5164-9_13

congestion spread to the entire network operating system the linkage effect is more outstanding, operational risk increased. Therefore, how to improve the stability, reliability, and risk control ability of the highway network to ensure its safe and efficient operation has become a prominent issue in the development of the highway while constantly strengthening the construction of the highway network and improving the capacity and efficiency of traffic flow.

From the point of the present study, most of the existing highway safety analysis method without considering the operating characteristics of highway into mesh distribution after risk for road network and traffic flow dynamic effect is still confined to sections and intersections as evaluation objects [1, 2], with accident statistics as evaluation index [3, 4], such as principal component analysis and analytic hierarchy process (AHP) method as the evaluation method of after evaluation [5, 6], multi-scale and multi-level representation is not from network real-time risk profile evolution.

In view of this, risks caused by non-traffic accidents and black spots will be included in the model for the first time in this paper, and TransCAD will be used to apply the model to risk analysis of real road networks.

## 13.2 Risk Analysis Model

On the basis of studying the existing risk analysis models and drawing lessons from the more common theoretical models in the world at present, this paper puts forward a more suitable grading model for regional dangerous goods road transport, in order to adapt to the lack of basic data in China. The model consists of three parts: risk of death, risk of vehicle loss, and risk of road loss.

The risk analysis model is:

$$D_r = \alpha D_{P,r} + +\beta D_{C,r} + \chi D_R, \tag{13.1}$$

$$D_{P,r} = P_r \times C_{P,S} \tag{13.2}$$

$$D_{C,r} = P_r \times C_{C,S} \tag{13.3}$$

$$D_{R,r} = P_r \times C_{R,S} \tag{13.4}$$

where $D_r$: risk value of section $r$; $D_{P,\,r}$: risk value of death in section $r$; $D_{C,\,r}$: risk value of vehicle loss in section $r$; $D_{R,\,r}$: risk value of road loss in section $r$; $P_r$: probability of dangerous goods accident in section $r$; $C_{P,\,S}$: cost of death caused by dangerous goods accidents caused by accident type $S$; $C_{C,\,S}$: vehicle loss cost caused by dangerous goods accident caused by accident type $S$; $C_{R,\,S}$: road repair costs caused by dangerous goods accidents caused by accident type $S$; $\alpha$, $\beta$, $\chi$ are 0 or 1.

According to the quantitative risk analysis model of dangerous goods transport, the dangerous goods transport accident rate $P_r$ includes the dangerous goods accident rate $P_{F,S}$ caused by accident type $S$. $P_{N,S}$, the accident rate of dangerous goods caused by non-traffic accidents; the accident rate of dangerous goods caused by accident black spots is composed of $P_{P,S}$ three parts.

$$P_r = P_{F,S} + P_{N,S} + P_{P,S} \tag{13.5}$$

where $P_{F,S}$ represents the probability of dangerous goods accidents caused by traffic accidents of accident type $S$ in this section. It is obtained by multiplying the national accident frequency and the average running time of section $r$ with the length of section $r$ and the accident probability of accident type $S$.

$$P_{F,S} = f_{F,S} \times P_S \times t_r \tag{13.6}$$

$$f_{F,f} = r_r \times L_r \times n_r \tag{13.7}$$

$$r_r = r_{0,r} \prod_{j=1}^{8} h_j \tag{13.8}$$

where $t_r$: the average running time of section $r$; $f_{F,f}$: accident frequency of section $r$; $r_r$: the expected frequency of road section $r$ accidents; $L_r$: length of road section $r$; $n_r$: the number of vehicles in section $r$; $r_{0,r}$: the national accident frequency; $h_j$: coefficient of regional enhancement or weakening.

The dangerous goods accidents caused by non-traffic accidents refer to the transportation accidents of dangerous goods caused by reasons other than traffic accidents. The typical dangerous goods accidents caused by non-traffic accidents include leakage of safety valve and leakage in the device. Overload of storage tank or failure of storage tank may lead to material leakage, thus leading to dangerous goods accidents. The failure of pressure relief valve and bursting disk will also lead to the occurrence of dangerous goods accidents under general operating conditions.

Generally speaking, the type of non-accidental leakage is largely related to the duration of use, so the failure rate is expressed by the time of use or the number of operations rather than the distance traveled. The accident rate of dangerous goods caused by non-traffic accidents $P_{N,S}$ was:

$$P_{N,s} = \eta_{tr} \tag{13.9}$$

$\eta$: the failure rate of single vehicle caused by non-accident; $tr$: the running time of dangerous goods transport vehicle in section $r$.

Section accident black spot in a long period of time, the frequency or number or characteristics of road traffic accidents are significantly more prominent than other normal road sections (or locations). It is mainly caused by the horizontal and vertical combination of roads, cross-sectional form, pavement structure form, traffic control mode, natural and climatic conditions, landscape performance, etc. The number of black spots and accident frequency was obtained by statistics. The model of dangerous goods accident rate of accident black spot is as follows:

$$P_{P,S} = \sum_{k=1} f_{k,r} \tag{13.10}$$

$f_{k,r}$: statistical accident frequency of the $k$th accident black spot in section $r$.

The death caused by road transportation accidents of dangerous goods can be divided into three categories: (1) on road death $N_{r,s}^{\text{in}}$, mainly passengers, is the direct victim of the accident consequences; (2) out of road deaths $N_{r,s}^{\text{off}}$, mainly residents and floating people along the line, the size of the impact range is related to the type of dangerous goods and transportation volume, and the probability of people appearing outdoors $P_r^{\text{out}}$ and the risk mitigation effect of indoor $a_r$ are also considered; (3) Personnel n of population gathering centers along the line $N_{r,s}^{\text{cent}}$, such as schools, hospitals, and shopping center. In the process of establishing the model, the human risk of population concentration center is included in the risk of off-road personnel. The establishment of death risk model is based on the number of deaths and the damage model of roads and vehicles. The model is as follows:

$$D_{P,r} = D_{P,r}^{\text{in}} + D_{P,r}^{\text{off}} \tag{13.11}$$

Among them: $D_{P,r}^{\text{in}}$: risk value of on road death in section $r$; $D_{P,r}^{\text{off}}$: risk value of death outside road in section $r$.

The risk model of on-the-road personnel death caused by dangerous goods accidents is calculated by multiplying the number of dead people in the road by the death cost of a single person as follows:

$$D_{P,r}^{\text{in}} = V_P \times N_{r,s}^{\text{in}} \tag{13.12}$$

Among them, $V_P$ is the death cost of a single person; $N_{r,s}^{\text{in}}$: the number of on road deaths caused by dangerous goods accidents caused by accident $S$ in section $r$;

When the dangerous goods accident occurs, all passengers within the radius of death will die. The number of passengers is determined by the number of vehicles within the radius of death and the average number of passengers. The model is as follows:

$$N_{r,s}^{\text{in}} = N_{r,C} \times K \tag{13.13}$$

$$N_{r,C} = 2Q_r \times R \times V \tag{13.14}$$

Among them: $N_{r,C}$: the number of vehicles within the death radius of dangerous goods accidents in section $r$; $K$: the average riding factor; $Q_r$: the total amount of dangerous goods transportation in section $r$; $R$: the death radius caused by the accidents of dangerous goods of unit mass; $V$: the traffic density of section $r$.

When a traffic accident occurs on the road, the accident point will affect the upstream, downstream and object traffic flow density. In order to more accurately express the change of traffic density in this process, the traffic density $V$ of section $r$ in the model is expressed as the sum of upstream, downstream and object traffic flow density after dangerous goods accident:

$$V = V_{\text{up}} + V_{\text{down}} + V_{\text{opp}} \tag{13.15}$$

Among them, $V_{\text{up}}$ is the upstream traffic density of the accident lane; $V_{\text{down}}$ is the downstream traffic flow density of the accident lane; $V_{\text{opp}}$ is the opposite traffic flow density of the accident lane.

There are two types of off-road deaths caused by dangerous goods accidents, one is the nearby residents and floating population, the other is the people in the population gathering center with relatively concentrated personnel such as schools, hospitals, and shopping centers. The risk model of off-road death caused by dangerous goods accident is as follows:

$$D_{P,S}^{\text{off}} = \left(N_{r,s}^{\text{off}} + N_{r,s}^{\text{cent}}\right) \times V_P \tag{13.16}$$

Among them, $D_{P,S}^{\text{off}}$ is the risk value of off-road death caused by dangerous goods accidents caused by accident $S$ in section $r$; $N_{r,s}^{\text{off}}$ is the number of off-road deaths caused by dangerous goods accidents caused by accident s in section $r$; $N_{r,s}^{\text{cent}}$ is the number of deaths in the off-road population gathering area caused by dangerous goods accidents caused by accident $S$ in section $r$.

The number of out of road deaths caused by dangerous goods accidents caused by the accident $S$ in section $r$ refers to the death toll within the death radius of dangerous goods accidents except the area outside the road, that is, the death toll within the influence area outside the road. The specific model is as follows:

$$N_{r,s}^{\text{off}} = \delta \times d_{p,r} \times \left(P_r^{\text{out}} + (1 - P_r^{\text{out}}) \times \alpha_r\right) \tag{13.17}$$

$$\delta = \left(\pi(Q_r \times R)^2 - 2Q_r \times R \times W_r\right) \tag{13.18}$$

where $\delta$ is the impact area outside the road; $d_{p,r}$: the population distribution density outside the road of section $r$; $P_r^{\text{out}}$: the probability of the affected personnel appearing outdoors in section $r$; $\alpha_r$: the risk mitigation coefficient of indoor personnel in section $r$; $W_r$: the road width of section $r$.

The risk model of vehicle damage caused by road transportation accident of dangerous goods is the product of the total number of vehicles within the death radius caused by the total amount of dangerous goods transportation accident and the average cost of each vehicle:

$$D_{C,S} = V_c \times N_{r,C} \tag{13.19}$$

The risk model of road damage caused by road transportation accidents of dangerous goods is the product of the road cost per unit length and the length of death diameter caused by the total amount of dangerous goods transported:

$$D_{R,S} = V_R \times 2Q_r \times R \tag{13.20}$$

Among them: $V_c$: unit vehicle cost; $V_R$: unit length road repair cost.

## 13.3 Conclusion

The risk value of a road section mainly depends on the accident rate of the road section and the personnel, vehicles, and roads in the consequence of the accident. Due to the different attributes of the road sections, the accident rates of the road sections calculated by the national average road accident rate often differ greatly. The distribution of the off-road population in the same district is the same, and the consequence of the risk of accident mainly depends on the population distribution on the road. That is, it depends on the value of traffic flow of each section.

**Acknowledgements** This work was supported by the National Key R&D Program of China (2017YFC0804807) and the Central Public Research Institutes Special Basic Research Foundation (2020-9067).

## References

1. Shifeng, N., Yongxiong, Z., Shadan, F.: Traffic safety evaluation method for highway section based on fault tree. J. Chongqing Jiaotong Univ. (Nat. Sci.) **32**(1), 87–90 (2013)
2. Jinhua, L., Li, L., Ning, Z.: Simulation-based evaluation of urban road intersection traffic safety. China Saf. Sci. J. **26**(12), 41–46 (2016)
3. Duqiang, X.: Evaluation on road traffic safety of the cities in Shaanxi province based on entropy—TOPSIS method. J. Saf. Sci. Technol. **11**(10), 116–120 (2015)
4. Xiaoli, H., Yaolong, L., Jin, D.: The assessment of road traffic safety risk based on grey relation and fuzzy comprehensive evaluation method. Math. Practice Theory **47**(7), 208–215 (2017)
5. Dawei, X., Junling, J., Zhu, B.: Study on traffic safety evaluation for freeway based on principal component analysis. J. Heilongjiang Inst. Technol. **28**(2), 46–49 (2014)
6. Guangzhen, Z., Xiaochun, J., Mingming, Z.: Evaluation for traffic safety of expressways based on improved analytical hierarchy process. Technol. Highw. Transp. **34**(3), 120–123 (2015)

# Chapter 14
# Virtual Reality Technology Facilitates Customized Clothing Design in C2M Business Model

**Sun Jian**

**Abstract** At the call of market demand, with the progress and development of science and technology, customized clothing design under C2M business model has become a breakthrough point for the upgrading and transformation of clothing industry in China. This paper discusses the effect of VR of artificial intelligence on this kind of clothing business model from three aspects of product manufacturing end, client end, and shopping experience under C2M clothing customization mode.

## 14.1 Introduction

With the development of the times, most consumers intend to increase their clothing pursuing on individualism, difference and exclusive sense, and focus on shopping experience. This kind of consumer demand brings more development chance to private customized clothing, and the new consumer-centric supply system has formed gradually and improved constantly. This clothing supply system completely subverts traditional producer-centered production-marketing mode, and zero inventory brought by customized product eradicates obstinacy caused by traditional producer-centered production-marketing mode. According to *An Analysis Report on Market Foresight and Investment Strategic Planning of China's Garment Design Industry*, with the consumer personalized, exclusive needs continue to strengthen, it promotes the clothing industry upgrading and transformation. In this process, customized production will become the inevitable trend of brand clothing development.

At present, most customized branded apparel is realized by C2M business model. C2M is the abbreviation of customer-to-manufacturer in English. It is a new business model of industrial Internet e-commerce [1]. With the support of the Internet, big data, and artificial intelligence technology, consumers can place orders directly to the garment factory through the clothing customization platform, and the factory directly designs and produces customized clothing according to the requirement of customers.

S. Jian (✉)
Dalian Polytechnic University, Dalian City, Liaoning Prov, China

 
L. C. Jain et al. (eds.), *Smart Communications, Intelligent Algorithms and Interactive Methods*, Smart Innovation, Systems and Technologies 257,
https://doi.org/10.1007/978-981-16-5164-9_14

In this process, VR plays a very important role in the two-way communication between customers and garment factory.

## 14.2 The Development of VR in the Field of Clothing

VR is a computer simulation system that can create and experience virtual world. It uses computer to generate a simulation environment. It is a system of multisource information fusion, interactive 3D dynamic scene, and entity behavior [2]. This technology integrates computer graphics, database design, control, fact distribution system, and multimedia technology [3]. It has strong human–computer interaction and will bring immersive feeling to the participants; at the same meanwhile, it has the multi-perception of vision, hearing, force sense, touch, movement and other human perception, even more, it has the imagination of building various environment and objective conditions in the virtual world.

From the end of the twentieth century to the beginning of the twenty-first century, with the research development and application of 3D human scanning technology and 3D fitting technology in the field of clothing, VR has increasingly made the clothing design, research development and exhibition sales rise to an unprecedented new stage. Through the 3D human body scanning system, we can get the accurate virtual 3D human body model of consumers. The 3D fitting technology can make the plane 2D template through virtual stitching, wearing on the human body model, so that both designers and consumers can see the effect of clothing fitting directly. The virtual fashion T-show displays the dynamic wearing effect of clothing in a realistic and vivid form, which not only shows the clothes but also saves the cost of renting real models and venues. In recent years, with the progress and improvement of virtual reality technology, wearing some VR devices can also enter a completely immersive virtual dressing experience. In such case, people and machines will have more intelligent human-computer interaction. Especially now, when the garment industry is upgraded to C2M customization mode, virtual reality technology provides strong support for the two-way communication and interaction between "manufacturing end" and "client end."

## 14.3 Virtual Reality Technology Facilitates C2M "Manufacturing End" Custom Clothing Design

### *14.3.1 More Vivid Model Display*

For the manufacturing end, the basis of customized clothing is the human body. By using 3D human body measurement technology and VR, we can set up full size of mannequins and various specifications of human body models in the computer. By

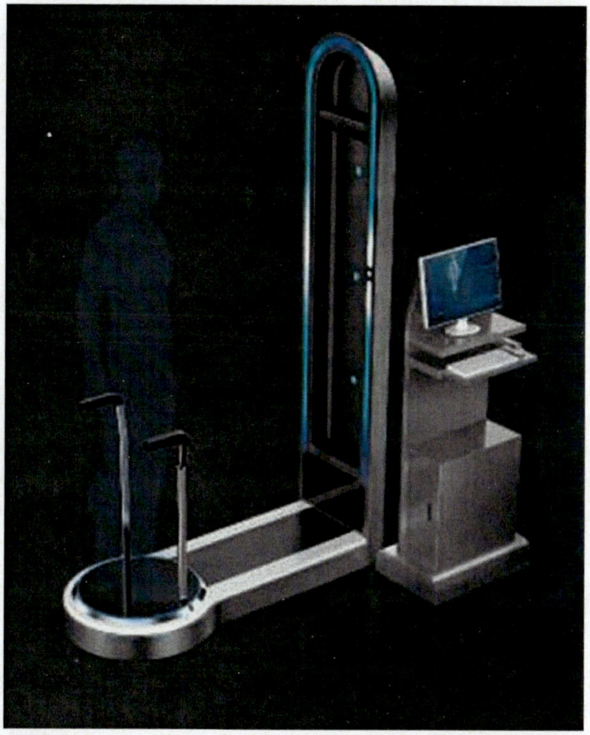

**Fig. 14.1** Bock technology's full-color intelligent 3D scanner

inputting and modifying the data of each part of the human body, we can adjust the body shape of the mannequin to make the appearance as close as possible to the real human body, which lays the foundation for the success of the fitting. The application of color high-definition scanner makes human body modeling more convenient and realistic. The color high-definition scanner is used to scan and model the human body's surface color, skin texture, and map automatically. As like as two peas, the 3D model of human body is exactly as same as the real person in facial features and hairstyles. This makes the later customized clothing design more suitable for the customer's figure, face shape, temperament, and other characteristics. For example, Shenzhen Boke technology's full-color intelligent 3D scanner, scanning time less than 60 s, using infrared camera and RGB camera as the scanning component, and using WebGL rendering technology, the established high-definition, realistic virtual human body model can be displayed on mobile app and PC (Figs. 14.1 and 14.2).

### *14.3.2 More Intuitive and Stereoscopic Appearance*

On the large-scale C2M clothing online intelligent customization platforms such as "Cotte" and "Yibangren," we can see many brand manufacturers are using virtual mannequins to display virtual clothing. Using VR, the plane patterns can be virtual

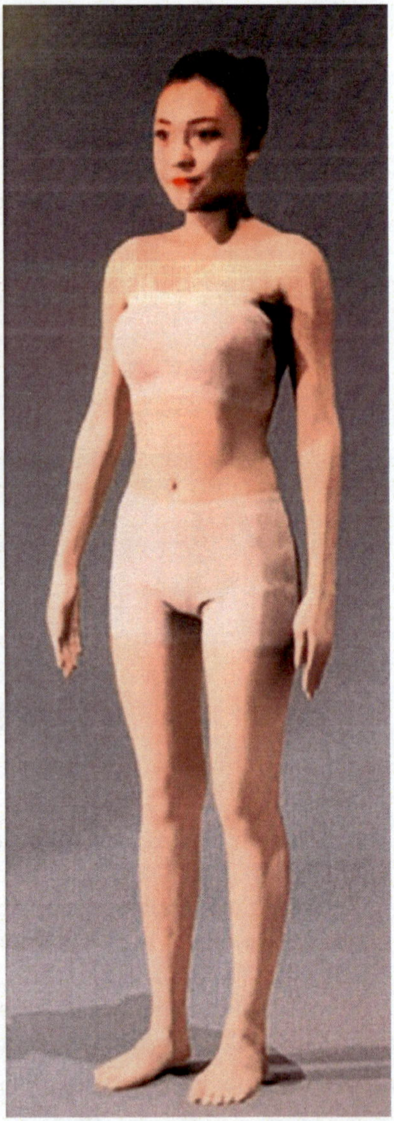

**Fig. 14.2** Human body model built by HD full-color scanner

stitched into 3D sample clothes, which can be worn on the virtual human body, and the human model can rotate 360 degrees to present the clothing effect more intuitively and stereoscopically. On the C2M customization platform, in addition to clothing manufacturers, there are also product links from fabric manufacturers. These fabrics can also be attached to the surface of 3D sample clothes with virtual reality technology. With the help of digital simulation technology, the fabric structure can be simulated interactively, showing various fabrics or soft or crisp different texture. At the same time, VR can also show the color and pattern of clothing vividly and

**Fig. 14.3** Visual and stereoscopic clothing effect presented by VR

stereoscopically in front of consumers and help consumers make purchase decisions in accordance with their wishes (Fig. 14.3).

### *14.3.3 Realize the Transformation from 2 to 3D in the Structure Design*

In the garment manufacturing end, the 2D pattern can be transformed into 3D clothing by using 3D visual virtual stitching technology in VR. Using this technology, we can completely simulate the real situation of sewing, making pieces up a complete clothing according to the sewing rules, and then wear it on the virtual model to see the effect. This can greatly reduce the number and times of real sample sewing, to improve the efficiency of garment production design and research development.

On the other hand, with the support of virtual reality technology, 3D garment can be transformed into 2D pattern. For example, there is a technology called "prototype stripping" in AGMS's CAD system of Japan, which can transform the 3D clothing into the plane prototype template, which is a very efficient method for making prototype template. Other 3D virtual clothing can also be transformed into 2D pattern. In this way, the channel of free transformation between 2 and 3D is opened in the field of garment pattern making, so that more efficient and accurate clothing structure design can be realized (Fig. 14.4).

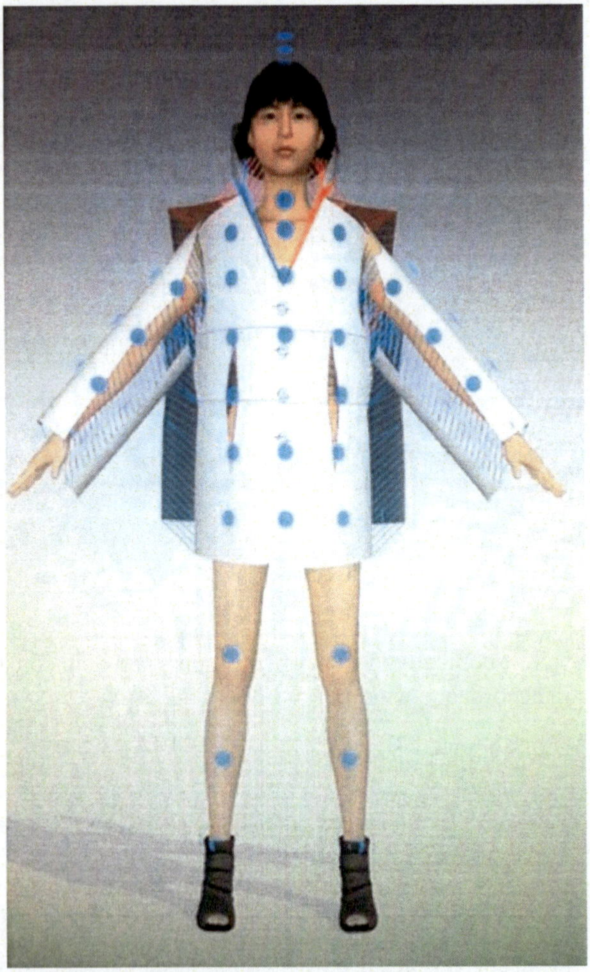

**Fig. 14.4** 3D visual virtual stitching technology

### *14.3.4 Facilitate Clothing Functional Research and Development*

Many customers have special requirements for customized clothing, which need clothing to adapt to some special body shape or have some functions. On the manufacturing side, intelligent VR provides technical support for the customization and research development of special clothing. For example, virtual human modeling technology can be used to simulate any special body shape and human body dynamics, which can be used as the basis of clothing design and pattern making to carry out customized research and development of special clothing and functional clothing. For example, the Barrier Free Clothing Research Center of Beijing Institute of Fashion Technology uses VR to simulate the daily dressing state of the disabled and develops a series of barrier free clothing accessories with both function and beauty, which

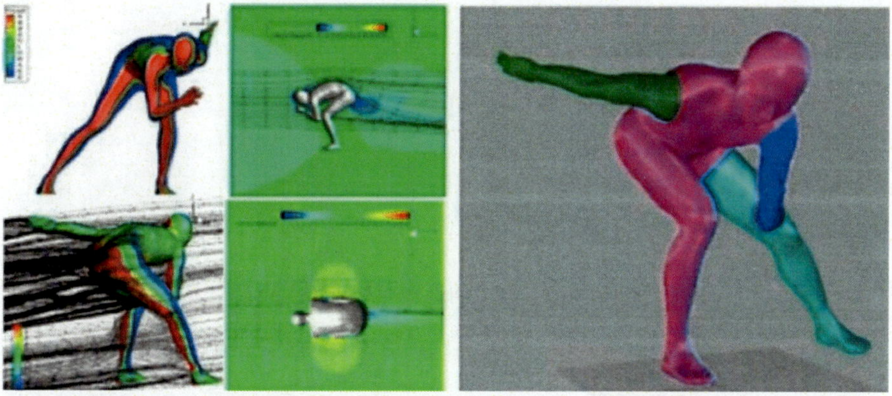

**Fig. 14.5** "Moving platform for wind tunnel evaluation" and "numerical simulation model for fluid dynamics"

meet the disabled people's requirement for easy to wear and take off, beautiful, and elegant as well. In the research development of high-performance clothing equipment for the Winter Olympic Games, the unit also used the VR to construct the "sports platform for wind tunnel evaluation" and "numerical simulation model of fluid dynamics," which provided important technical support for the research development and customization of competition clothing related to the Winter Olympics (Fig. 14.5).

## 14.4 VR Facilitates C2M "Client End" Custom Redesign

At present, in many C2M online intelligent customization platform, customers can "redesign" products through human–computer interaction. The field of redesign includes changing fabric, profile, and design details. We can also put forward more exclusive design requirements, such as embroidering one's favorite proverbs on clothes. With the support of VR, customers' choices will be immediately displayed on the virtual models in the platform interface in three dimensions, so that customers can see the results of their choices intuitively and enhance the sense of participation of consumers in fashion design. This kind of "redesign" of products with full participation of customers corrects the subjective assumption of clients' needs at the manufacturing end, more respects the personalized needs, and enhances the specificity of customized products. For example, on the "Cloud Clothing Customization" intelligent clothing customization platform developed by Shenzhen Bock Technology, customers can directly place orders online after completing the redesign of a garment from style to fabric through the Internet. After receiving the order, the factory will take 5–20 days to complete the production of customized products according to different categories and then send them to consumers (Fig. 14.6).

**Fig. 14.6** In the "Cloud Clothing Customization" platform, the fabrics selected by customers are presented on virtual ready-made clothes

## 14.5 VR Makes Shopping Experience More Interesting Under C2M Mode

With the support of AI technology such as VR, customers' shopping experience on C2M online intelligent customization platform is full of fun of human–computer interaction. For example, in Dayang Chuangshi's intelligent clothing customization platform, the ruler measuring the human body is equipped with a chip. The measurement data in the real situation can be immediately uploaded to the customized garment size system of the intelligent platform, which makes the customization on the Internet more convenient and efficient.

VR also helps consumers to experience the quality of clothing products. For example, in 2016, in order to let consumers have a more intuitive understanding of an innovative technology fabric, Nike introduced the VR experience device in Shanghai stores for the first time [4]. In the virtual world, through the form of games, the consumers wearing new products can exercise in different forms greatly, feel the comfort of clothing in the sports state, and let consumers experience the excellent performance of products in the virtual game. This novel and interesting interactive experience also has a positive impact on customers' purchase decisions (Fig. 14.7).

## 14.6 Closing Remarks

At the recent 2020 world VR conference, people from all walks of life agreed that VR will be the next outlet of information industry. As Ren Aiguang, deputy director

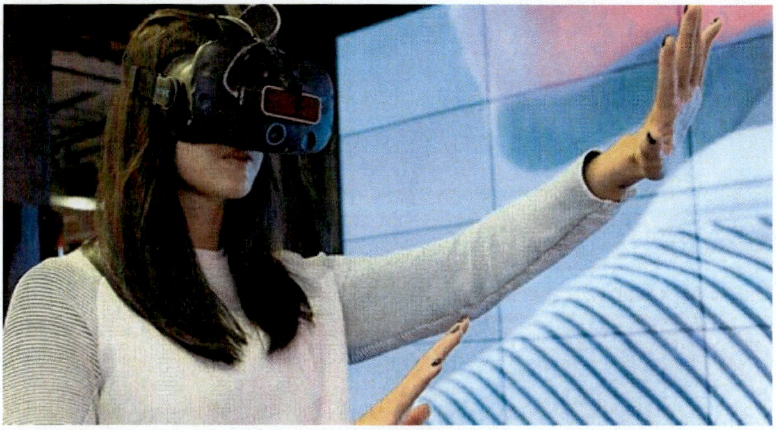

**Fig. 14.7** Wearing VR experience device to test clothing

of China Electronic Information Department, said: "at this stage, VR is increasingly improved, the market scale is expanding, and various VR + applications are also accelerating the landing." According to the prediction of CCID consultants, the scale of China's VR/AR market will exceed 100 billion yuan in 2022, becoming the growth center of global virtual reality market [5].

Currently, VR and related industries are strongly supporting the upgrading and transformation of China's clothing industry. As far as C2M garment customization system is concerned, from the perspects of the manufacturing end, client side and shopping experience of the product, the combined application of VR and other high-tech makes this business model centered on customer demand come true. With the continuous improvement and development of VR, it is expected that in a more 3D, more realistic, more intelligent, and more interesting virtual world, people's pursuit of clothing beauty will leave more brand-new marks of the times.

## References

1. Jianjian, L., Shuhua, B.: A study and discussion on the development of C2M E-commerce model-take coolt intelligence as an example. Electron. Commer. **12**(15), 24–25 (2019)
2. Yilin, L.: Application and research of virtual reality technology in garment design and industrial plate making. Textile Report **12**(20), 53–54 (2019)
3. Linghui, S.: Research on the application of virtual reality technology in B2C platform based on SNS_taking clothing accessories as an example. Office Autom. **6**(15), 48–50 (2020)
4. Yiyun, L., Zhi-wei, J.: Application of virtual reality technology in clothing marketing. Text. Technol. Progress **7**, 53–55 (2020)
5. Lexuan, Z.: 2020 world VR industry conference Yunfeng opening in Nanchang, Jiangxi. People's Post **10**(20), 1 (2020)

# Chapter 15
# Research on Identification Model of Special Transformer Stealing Electricity Based on Vector Similarity Matching Algorithm

**Xin Xuan, Yazhuang Cao, Shuming Wang, and Tingyan Jiang**

**Abstract** In today's society, the act of stealing electricity not only loses a large amount of electricity fee income, but also brings property losses to the state grid, and at the same time, it is extremely easy to lay hidden security risks. The characteristics of the means of stealing electric power have shown the development from the private construction to the specialization, concealment and networking. Internet big data, this paper "precise marketing" thinking, according to the special transformer users power samples form special transformer power fingerprint, and special transformer users of electricity characteristic signal form of special transformer power use fingerprint vector similarity comparison, summed up the special transformer power matching algorithm based on vector similarity identification model. In practice, the algorithm is efficient, simple and accurate, and can effectively assist the inspectors to identify electric larceny.

## 15.1 Introduction

With the development of State Grid Information Construction and the popularization of smart meters, a large number of users' real-time data such as voltage, current and power have been accumulated. How to accurately identify and locate the suspected power stealing users through the abnormal power consumption signals in the user's electrical signal data is a research hotspot at home and abroad [1, 2]. At present, a large number of professionals has carried out research on electricity theft identification, but there are still some problems such as low practicability, low accuracy and low-calculation efficiency. Based on this, the research on the identification model of electric larceny of special transformer based on vector similarity matching algorithm proposed in this paper can be used to establish the power stealing identification model of special transformer based on the existing multi-source data of marketing specialty by applying the method of big data mining. By using the identification model of

X. Xuan · Y. Cao (✉) · S. Wang · T. Jiang
Beijing CHINA-POWER Information Technology Co., Ltd. State Grid Information & Telecommunication Group, Beijing 100085, China
e-mail: cyaz011@163.com

 
L. C. Jain et al. (eds.), *Smart Communications, Intelligent Algorithms and Interactive Methods*, Smart Innovation, Systems and Technologies 257,
https://doi.org/10.1007/978-981-16-5164-9_15

transformer stealing electricity, the abnormal situation of voltage, current and power of users can be monitored, and various means of stealing electricity such as private capacity increasing and bypassing measurement can be identified, and the suspected power stealing users can be found.

## 15.2 Analysis of Power Stealing of Special Transformer

### *15.2.1 Summary of Power Stealing Methods for Special Transformer*

There are many ways to steal electricity from special transformer users. The essence of stealing electricity is to avoid measurement and reduce measurement by changing the metering device. The main idea of stealing electricity is to start from the formula of electric quantity measurement: as shown in formula 15.1, the power stealing user can change the parameters such as voltage, current and power factor to achieve the goal of no or less measurement of electric quantity and complete the theft In addition, some electricity stealing users steal electricity through direct off meter wiring to avoid measurement and special transformer capacity increase. To sum up, the power stealing methods of special transformer users are summarized as follows: (1) voltage loss method, (2) under current method, (3) differential expansion method, (4) phase shift method and (5) no meter method [3].

$$Q = U \times I \times \cos\varphi \times T \tag{15.1}$$

Formula 15.1: electricity metering formula.

Note: $Q$—power consumption, $U$—voltage, $I$—current, cos $\varphi$—power factor, $T$—time.

### *15.2.2 Summary of Power Consumption Characteristic Signals of Special Transformer Users*

At present, a provincial power company of State Grid can collect the electricity consumption signals of special transformer users, such as electric quantity, voltage, current, power, power factor, etc., and the acquisition frequency is once per hour. Using big data thinking, through the electric quantity interface table, voltage interface table, current interface table and power interface table, these characteristic signals of electricity consumption are summarized to form a "power consumption profile," and the judgment conditions of each feature vector are used to scan the "power consumption fingerprint database," which can further effectively judge whether the user has abnormal electricity consumption behavior.

The characteristics of normal electrical signals of users are as follows:

(1) Voltage: the voltage value and the power supply voltage value are basically equal, and the value remains basically unchanged.
(2) Current: the current value changes with the change of external load, but the three-phase current basically keeps balance.
(3) Power factor: the power factor value is basically stable, and the phase angle is stable.
(4) Line loss: the line loss value will fluctuate within a certain range, generally less than 7%.

The power, voltage, current, power, power factor and other electrical signals of special transformer users will have certain changes in their daily use. This change has nothing to do with whether the user has electricity stealing behavior. However, if the change of power consumption signal exceeds a certain range of values, we can judge it as abnormal power consumption and form a "power fingerprint database." Through further analysis, we can judge the power consumption abnormality know whether it has stealing electricity.

## 15.3 Current Situation of Electricity Stealing Samples of Special Transformer Users

In many references [4–7], the traditional data mining methods are described in detail, and the process basically follows the crsip-dm process. However, according to the current situation of Zhejiang electric power company of State Grid, there are few electricity stealing samples of special transformer users. If the traditional data mining algorithm training model is used to identify the power stealing of special transformer users, there will be some inevitable problems, such as the construction of the training set, the model's recall and precision rate.

Therefore, the idea of "user portrait" based on the "precision marketing" thinking of Internet big data [8, 9] and the idea of "user portrait" based on the characteristics of certain users' online shopping behavior can also be applied in the field of anti-power stealing. Through the analysis of multi-source data of marketing specialty, this paper extracts the characteristic vector and vector threshold of electricity stealing behavior of national historical electricity stealing users, describes the "user portrait" and "stealing fingerprint" of suspected electricity stealing, realizes the accurate positioning of suspected users, and forms the fingerprint database of stealing electricity.

Due to the small number of electricity stealing samples of special transformer, it is particularly important to summarize the electricity consumption characteristics of special transformer users for small sample analysis. By summarizing the existing electricity stealing samples, the power stealing fingerprints of special transformer users are formed, as shown in Fig. 15.1.

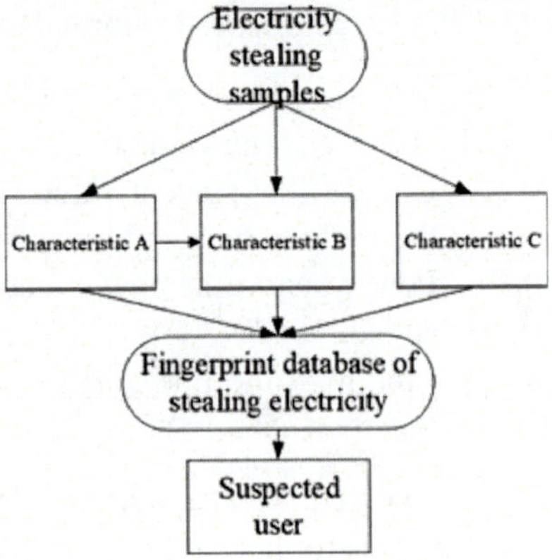


**Fig. 15.1** Identification process of stealing electricity based on small sample

## 15.4 Power Stealing Identification of Special Transformer Based on Vector Similarity Matching Algorithm

### *15.4.1 Vector Similarity Comparison Method*

The definition of similarity measure is to compare the degree of similarity between individuals. If the value of similarity measure is larger, the similarity between individuals will be greater; otherwise, it will be opposite. In a two-dimensional space right angle, the meaning of cosine similarity is: the similarity between two vectors $a(x_1, y_1)$ and $b(x_2, y_2)$ is expressed by the cosine value of the angle between vectors. When the cosine value of the included angle is closer to the value of 1, the angle between the two vectors is closer to 0 degree, which means that the two vectors are more similar, as shown in Fig. 15.2.

The included cosine formula of any pair of vectors in two-dimensional space is shown in Formula 15.2.

$$\cos(\theta) = \frac{a \cdot b}{|a| * |b|} = \frac{(x_1, y_1) \cdot (x_2, y_2)}{\sqrt{x_1^2 + y_1^2} \times \sqrt{x_2^2 + y_2^2}} = \frac{x_1x_2 + y_1y_2}{\sqrt{x_1^2 + y_1^2} \times \sqrt{x_2^2 + y_2^2}} \quad (15.2)$$

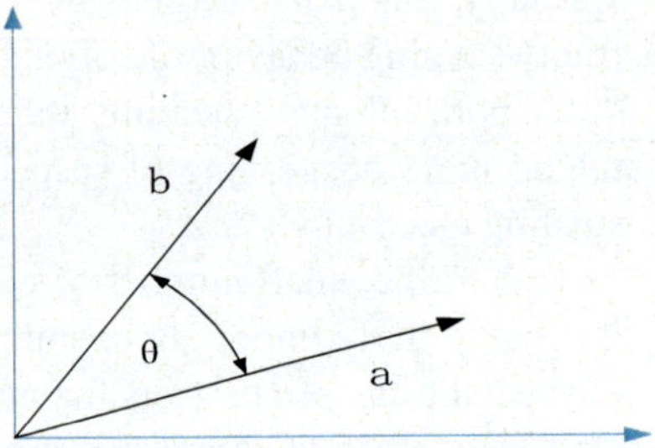


**Fig. 15.2** Comparison of similarities between vectors a and b

Formula 15.2: Formula for calculating the included cosine of two-dimensional space vector.

Note: $a$ and $b$ represent two different vectors, $(x_1, y_1)$ and $(x_2, y_2)$ are the coordinates of vector $a$ and $b$, respectively.

The $n$-dimensional space vector can be inferred from the two-dimensional space vector. Assuming that $a$ and $b$ are two vectors in $n$-dimensional space, the cosine formula for the angle between alpha and beta can be obtained as shown in Formula 15.3.

$$\cos(\theta) = \frac{a \cdot b}{|a| * |b|} = \frac{\sum_{i=1}^{n}(x_i \times y_i)}{\sqrt{\sum_{i=1}^{n}\left(x_i^2\right)} \times \sqrt{\sum_{i=1}^{n}\left(y_i^2\right)}} \tag{15.3}$$

Formula 15.3: Formula for calculating included cosine of n-dimensional space vector.

Note: $a$ and $b$ are two vectors in $n$-dimensional space, $(x_1, \ldots, x_n)$ and $(y_1, \ldots, y_n)$ are the coordinates of vector $a$ and $b$, respectively.

### 15.4.2 Identification of Power Theft by Special Transformer Users

Summarize the samples of power theft in history, form the special transformer user power theft fingerprint library, set it as the reference standard, and transform the user power signal to be detected into the power vector with the same dimension as the special transformer power theft fingerprint library to detect whether the user has power theft at a certain time. The specific identification process is shown in Fig. 15.3. Read several electric characteristic vectors in electric quantity interface table, voltage interface table, current interface table and power interface table to form user's electric picture. Scan user's electric picture into special transformer's electric fingerprint database by threshold value of electric anomaly signal index of special transformer user, form special transformer's electric fingerprint database and compare it with special transformer's electric stealing fingerprint database for vector similarity. Match, complete the algorithm and output the list of suspected power theft from special transformers.

### 15.4.3 Algorithmic Process Analysis

By matching the vector similarity between the "power fingerprint library" and the "power theft fingerprint library," the power user can be judged whether there is power theft. The special transformer power theft identification process based on the vector similarity matching algorithm is as follows:

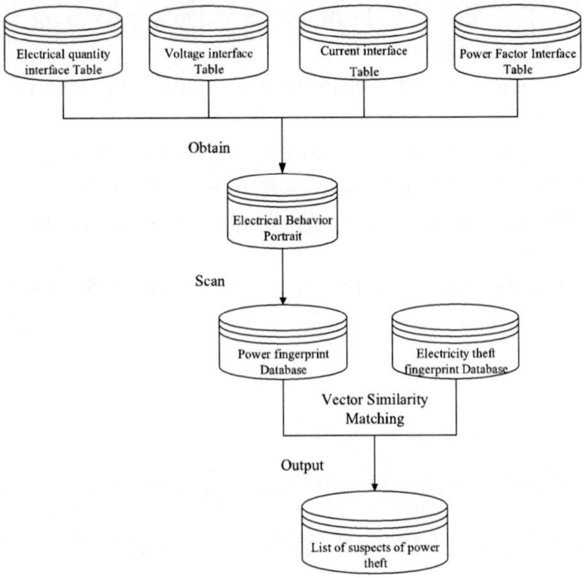


**Fig. 15.3** Identification of power theft by dedicated transformer users based on vector similarity

(1) Summarize the existing sample of power theft by users and form special transformer power theft fingerprint library;
(2) Collect the user's power consumption data and form the user's power consumption picture;
(3) Construct user's electric picture into special transformer power fingerprint database by threshold value of user's electric abnormal signal index of special transformer;
(4) Carry out vector similarity calculation between special transformer power fingerprint database and special transformer power theft fingerprint database;
(5) Output the result to the suspect scoring table to form the suspect list of power theft of special transformer.

### 15.4.4 Application Results of Algorithmic Model

Through field verification in Zhejiang Province, it is shown that the special transformer power theft identification model based on vector similarity matching algorithm can detect abnormal power consumption behaviors of users in time and accurately identify power theft users. The algorithm is efficient, concise and accurate, and can effectively assist inspectors in identifying power theft.

## 15.5 Conclusion

The identification model of special transformer power theft based on vector similarity matching algorithm can effectively solve the problems of less samples of power theft from special transformer users. At the same time, the algorithm is simple and easy to understand, which can effectively improve the calculation efficiency and facilitate operator to understand and optimize parameters. The following research directions can be carried out: (1) adding feature components and describing power images in detail; (2) optimizing threshold parameters of users' power anomaly signal indicators; (3) expanding special transformer power theft fingerprint library to enhance the recognition of different power theft methods by algorithm. Improve the convenience and accuracy of model identification for power theft.

## References

1. Yadi, T.: Research on anti-theft technology application of electric power information acquisition system. Mater. Decoration **12**(42), 211–212 (2016)
2. Bo, L., Ning, G., Hongmiao, C.: Anti-anti-theft scheme for power supply enterprises. Rural Electrician **26**(4), 48 (2018)
3. Zhongjun, M., Yunruo, X.: Discussion on electricity theft and anti-electricity theft in new form. Commun. World **25**(5), 244–246 (2018)
4. Junxing, X., Chuan, L., Yingna, L.: Electricity theft analysis based on BP neural network optimized by genetic algorithm. Software **38**(11), 18–23 (2017)
5. Hu, Z.: Cluster analysis of electric energy metering data and research on detection of power theft. China (2017)
6. Chuan, L.: Intelligent Clustering Analysis and its Application. Science press, China (2016)
7. Salini, B., Deepti, M.: Applying CHAID algorithm to investigate critical attributes of secured interoperable health data exchange. Int. J. Electron. Healthc. **8**(1), 25–50 (2015)
8. Hongyan, Z.: Discussion on application of power supply marketing service based on big data. Electromech. Inf. **18**(15), 169–171 (2018)
9. Zeying, H., Jiachen, X.: Precision marketing of big data—a tool for efficient marketing. Modern Mark. (Operating Version) **24**(6), 101 (2016)

# Chapter 16
# International Checkers Evaluation Algorithm Based on Evolutionary Neural Network

**Zhenghao Pan, ShuQin Li, and Shijing Gao**

**Abstract** As a chess species with a long history, research on drafts is mostly limited to the improvement of the chessboard representation method and the static evaluation function. Due to the limitations of the static evaluation function itself, there are problems such as insufficient evaluation accuracy and low judgment accuracy. This paper summarizes and analyzes the existing checkers evaluation algorithm and proposes an evolutionary neural network-based checkers evaluation algorithm. First, the neural network is used to train on different datasets, and the models trained on different datasets are selected as the parent. The model network, through the game between the parent and the offspring, selects the excellent evaluation neural network to keep as the next generation of the parent and iterates continuously and improves the accuracy of the situation evaluation by controlling the step length of the new mutation parameter of the search neural network, thereby get a good position evaluation function. Experiments show that its effect is better than the method based on static situation evaluation.

## 16.1 Introduction

International checkers [1], as an ancient chess with a long history, simple rules and various moves, is loved by many people in the world. International checkers 8 × 8 international checkers, as shown in Fig. 16.1, each side holds 12 pieces, which are placed in four rows near the baseline. Each chess piece can only walk one square in the left front or right front direction. Chess pieces that meet the conditions of "eating pieces" must undergo "eating pieces." When one's own chess piece reaches and can stay at the opponent's bottom line, it can complete the "becoming to a king"

Z. Pan (✉) · S. Li · S. Gao
School of Computer, Beijing Information and Science and Technology University, Beijing 100101, China
e-mail: lishuqin_de@126.com

S. Li · S. Gao
Sensing and Computational Intelligence Joint Lab, Beijing Information and Science and Technology University, Beijing 100101, China

 
L. C. Jain et al. (eds.), *Smart Communications, Intelligent Algorithms and Interactive Methods*, Smart Innovation, Systems and Technologies 257,
https://doi.org/10.1007/978-981-16-5164-9_16

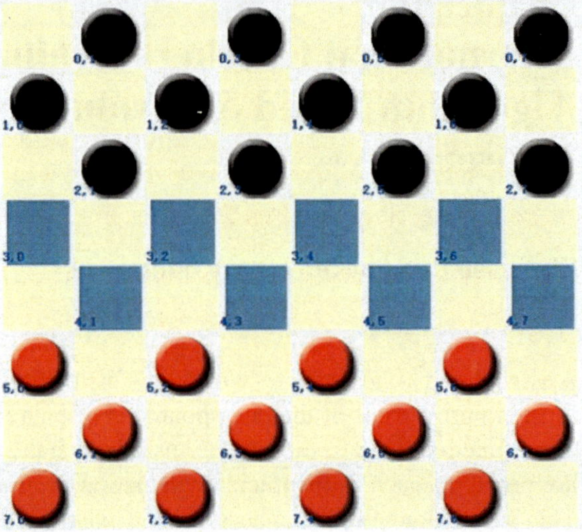


**Fig. 16.1** 8 × 8 checkers board

operation. After becoming a king chess, it can move any number of grids in the four directions of left front, right front, left back, and right back to walk. When one party's chess piece is empty or unable to move, it is judged to be negative. After one party applies for summation, the other party can choose to agree or disagree. If it agrees, draw chess as the final result; in case of opposition, the opposing party must win in the next 40 parts, otherwise it will be judged as negative.

Computer game is considered as one of the most challenging research directions in the field of artificial intelligence, attracting more scholars to study it. It can be said that the evaluation algorithm is very important. The so-called chess evaluation function [2] is the evaluation and quantification (scoring) of the current chessboard situation, so as to decide whether to choose or not according to its value in the search process. Evaluation function is the core of strategy in computer game system. Therefore, it can be seen that the evaluation function is used to judge whether the whole game is good or bad, and then affects the final game.

The value evaluation function of traditional algorithm is mostly static evaluation function [3–6]. The so-called static evaluation function, that is, after the function has passed the formatted regulations, it only changes its output value by changing the relevant parameters, so as to achieve the evaluation effect of the current situation. Traditional algorithm can realize basic man-machine game, but its algorithm still has many disadvantages, including.

The defect of judging the direction of the situation form by the value evaluation function: Because the variables of the evaluation function are only a limited number of situation characteristics, it may lead to the deviation of the determined chessboard value, thus affecting the situation selection of the search algorithm.

Defects of the value evaluation function in the direction of situation judgment accuracy: In essence, the static evaluation function is still a linear regression problem, which may lead to a great error precision of the model fitting effect, thus leading to a large error in the evaluation score of the current situation.

In recent years, from deep blue to AlphaGo [7] to AlphaZero [8], the game program of chess is becoming more and more perfect, and the static evaluation function is replaced by deep learning neural network, which has achieved good results. Aiming at the problem that static evaluation function cannot accurately judge the value of the current situation, this paper studies the situation value evaluation by using evolutionary neural network.

## 16.2 Design Idea of International Checkers Evaluation Algorithm Based on Evolutionary Neural Network

Because artificial neural networks are mostly mathematical models that can be understood by computers, we can get our own judgments by transmitting the input values forward and compare our own judgments with the real values to transmit the errors back to update our own network parameters. In neural network, the gradient descent method points out the optimal direction and keeps changing in this direction. However, the simple gradient descent method may fall into the local optimal situation, resulting in the obtained value not being the optimal value. Therefore, this paper introduces evolutionary neural network, which uses the original points to create new points, determines the starting point of the next generation by selecting, fixes the structure of the parent neural network, and uses it to generate multiple children with the same structure but slightly different connection strength between networks. After scoring through the game between parents and children, each child network is given a score value. The children are combined into a neural network in proportion, which is used as the parent of the next iteration, and then iterates continuously. Compared with gradient descent neural network, it has some shortcomings in updating speed, but it can better avoid falling into local optimum. Therefore, evolutionary neural network algorithm can be parallel, which can greatly reduce the training time.

The system designed in this paper is mainly composed of a deep neural network pre-model and an evolutionary neural network model. The overall system framework is shown in Fig. 16.2.

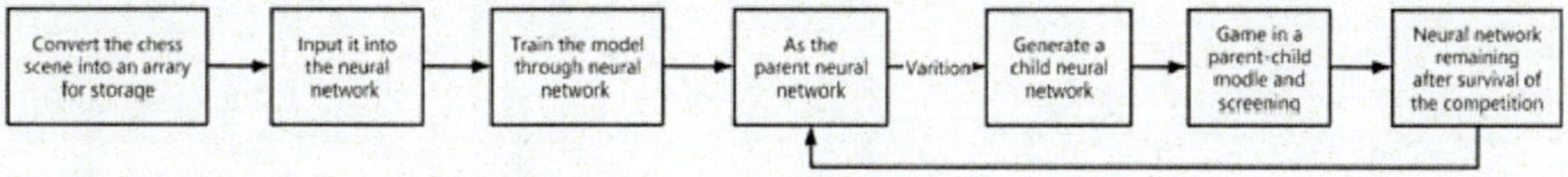


**Fig. 16.2** Overall design framework

## 16.3 Implementation of International Checkers Evaluation Algorithm Based on Evolutionary Deep Learning

### *16.3.1 Deep Neural Network Pre-model Design and Implementation*

According to about 20,000 chessboards recorded in international checkers competitions over the years collected on the network [9], the data are cleaned; that is, the words irrelevant to the chess playing process in the chessboard are deleted, only the data of both sides' players are kept, and the chessboard is restored according to the corresponding recording rules in the chessboard, and the state of the chessboard after each player is saved into the form of chessboard matrix until the end of the game. At the same time, the situation of the game will be marked, with victory expressed by 1 and failure expressed by $-1$.

The cleaned chessboard data are stored as an $8 \times 8$ matrix as the input of neural network, and the output layer of neural network contains a neuron and outputs a number between $[-1, +1]$, where the closer the number is to $-1$, the higher the opponent's winning probability and the greater the possibility of our losing; On the contrary, the number is closer to $+1$, and the higher our winning rate is, it reflects the assessment of the current situation and whether it is beneficial to our own side. In this paper, a three-layer fully connected neural network is designed, and the network structure is shown in Fig. 16.3.

At the same time, the parameters related to neural network are set, the training times are 300 times, and the learning rate is 0.05, so as to avoid the occurrence of local optimum as much as possible.

At the same time, Adam's gradient descent method is used to optimize the error. The principle of Adam's gradient descent algorithm is to make an exponential weighted average of the original gradient, and then update the weight after normalization

**Fig. 16.3** Pre-model structure of deep neural network

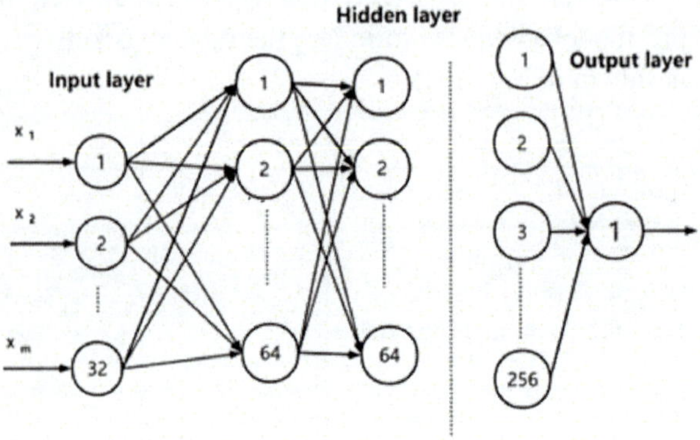

$$\begin{cases} v = \beta_1 v + (1 - \beta_1) dw \\ s = \beta_2 s + (1 - \beta_2) dw^2 \\ w = w - \alpha \frac{v}{\sqrt{s + \varepsilon}} \end{cases} \tag{16.1}$$

where $\beta_1 = 0.9$, $\beta_2 = 0.999$, $\varepsilon = 10^{-8}$; among them, $dw$ is the original gradient calculated by us, $V$ is the gradient calculated by exponential weighted average, and $S$ is a smoothing of the square of the gradient.

At the same time, the tanh function is regarded as the active function, because the range of its function is $[-1, +1]$, which is in line with the output value of the network we are involved in.

After several rounds of training, the error rate is finally controlled at 0.17, which can initially meet our fitting requirements and pave the way for the evolutionary neural network behind us.

### 16.3.2 Design and Implementation of Evolutionary Neural Network Mode

In this paper, the static evaluation function cannot accurately judge the value of the current situation, so the neural network is used as the value evaluation function instead. Therefore, firstly, neural networks are used to train in different datasets, and finally 10 well-trained models on different datasets are selected as parent model networks, and each neural network has a unique feature vector $\sigma_i$. It is determined by the weight and offset of the current neural network and the value of the currently set chess. Each neural network obtained is iterated, and the step size for searching new mutation parameters of the neural network is controlled by $\sigma_i$.

Based on the variation idea of evolutionary neural network and the change of control parameters through $\sigma_i$, a new child neural network is generated by adding or deleting connections, calculating the sum of the weights and offset numbers in the new neural network as $N$ again and updating it to $\sigma_i'$. And, each neural network was selected to play games with the rest of the networks. If one was won, the score would be 20 points, the score would be even, 10 points, and the score would be negative, 0 point. After five cycles, the 10 models with the highest scores were selected as the parents, and competitive screening was conducted again, finally the best optimized model was selected as our evaluation function.

When the model is implemented, firstly, the values of chess pieces are assigned, which are $+1$ for own common chess pieces, $-1$ for enemy common chess pieces, +K for own king chess, and −K for enemy king chess. At the same time, the weight of neural network and the evolutionary adaptive value of bias function are set to 0.05 and limited to $[-0.5, 0.5]$, which is used to control the step size of offspring evolution The initial value of k is set to 2, and it is limited to [1.0, 3.0], and the prescribed iterative function [10] is:

$$\sigma_i(j) = \sigma_i(j)\exp(\lambda N(0, 1)) \quad i = 1, 2, \ldots 10, \quad j = 1, 2, \ldots N \tag{16.2}$$

$$w_i'(j) = \sigma_i(j) + \sigma i'(j)N_i(0, 1) \quad i = 1, 2, ..10, \quad j = 1, 2, \ldots N \tag{16.3}$$

where $N$ is the sum of the number of weights and the number of offsets in the neural network, at the same time

$$\lambda = 1/\sqrt{2\sqrt{N}} \tag{16.4}$$

sum of the number of weights and the number of offsets of the ith neural network. $N_i$ is the Gaussian distribution between (0, 1).

At the same time is used to update the value of king [11].

$$K_i' = K_i \exp\left(\left(1/\sqrt{2}\right)\right) N(0, 1) \tag{16.5}$$

The value of the same situation will be different through the renewal of the offspring. Then, through the circular game between the parent and the offspring, the more advanced strategy will be chosen to keep it, and the iteration will continue, finally forming a nearly perfect evaluation function.

## 16.4 Experimental Results and Analysis

In this paper, the proposed algorithm based on evolutionary neural network is compared with the traditional game algorithm based on static evaluation function in the literature [6]. In the experimental process, the search algorithm uses alpha-beta pruning algorithm, and the search depth is set to six layers. The experiment is divided into two batches, each batch has 100 games, and every 10 games has a winning rate statistics. It can be seen that when black pieces use neural network evaluation function, while white pieces use the static evaluation function to play games. The statistical results of white pieces winning are shown in Fig. 16.4. It can be seen that the winning rate eventually approaches about 70%.

The second batch of black pieces played games with the static evaluation function and white pieces played chess with the evaluation function trained by the neural network in this paper. The result of statistical analysis of the victory of white pieces is shown in Fig. 16.5. It could be seen that the victory rate of white pieces approached 65% when using neural network to evaluate the function.

Two batches of experimental results show that the proposed evolutionary neural network algorithm is superior to the traditional situation assessment algorithm.

Fig. 16.4 Loss rate diagram

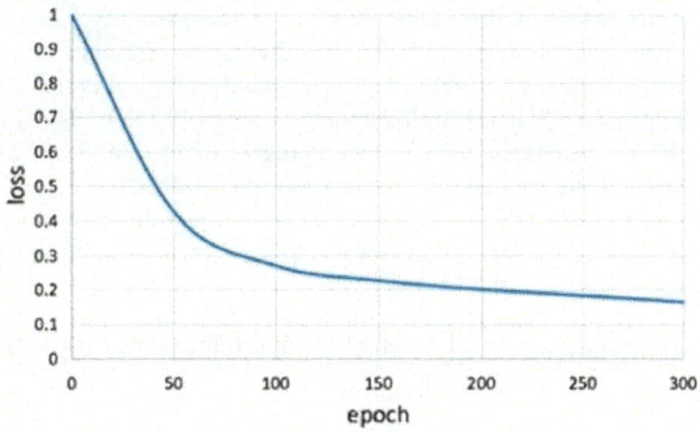


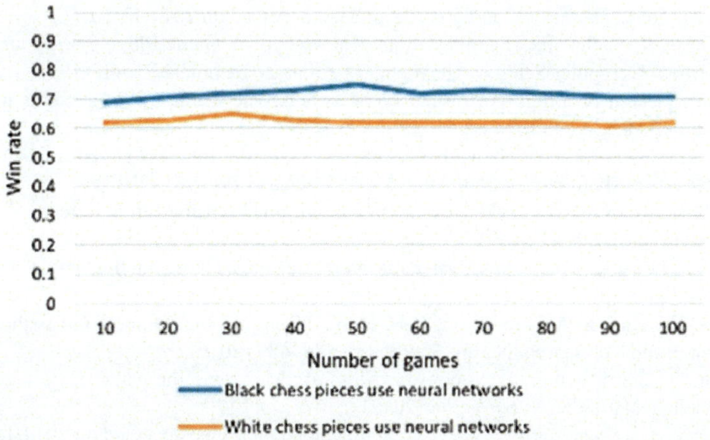


Fig. 16.5 Winning results of black pieces and white pieces using the evaluation function proposed in this paper

## 16.5 Conclusion and Prospect

The value evaluation function of a game program is a benchmark to evaluate the quality of a game program. A game program based on the static evaluation function cannot improve its overall accuracy and ability because of the defects of the evaluation function and its lack of learning ability. The game program based on deep learning mentioned in this paper can generate a neural network pre-model based on a large number of datasets. At the same time, through the evolution of the network through self-game, it constantly perfects the deficiencies, and finally improves the game level of the program.

As for future programs, we can combine deep learning with intensive learning, or realize self-game update like alpha-zero. Next, we will constantly debug the relevant parameters of neural network and carry out reinforcement learning and fusion.

**Acknowledgements** This work is supported by 2020 Promote Connotation Development of Universities-Scientific Research Training Program for College Students of BISTU(No.5102010805), by Key potential projects of Promoting Research Level program at Beijing Information Science and Technology University. (NO. 5212010937), by Normal projects of General Science and Technology research program (NO. KM201911232002), and by Construction Project of computer technology specialty (NO.5112011019).

## References

1. Chinese Association for Artififcial Intelligence. Chinese Association for Artififcial Intelligence
2. http://computergames.caai.cn/jsgz17.html. Last accessed on 2014/12/10
3. Andreas , J.: Schaeffer, Search Vesus: Knowledge in Game-Playing Program Revisited. Deperment Computer Science, University of Albama, Technical Report (1998)
4. Wu, Y.: Design of a high-performance checkers engine. Comput. Prog. Skills Maintenance (02), 63–67+82CNKI (2014)
5. Su, Z., Li, .S.Q., Lu, J.Y., Zheng, L.W., Fan, S.C.: The realization of genetic algorithm in terms of checkers evaluation function. Appl. Mech. Mater. 1979–1985 (2013)
6. An, M.M., Li, S.Q.: A checkers evaluation algorithm. J Beijing Inf. Sci. Technol. Univ. (Nat. Sci. Edn.) **32**(02), 85–88 (2017)
7. Gui, Y.Y.: Research on a game system of international checkers. Intell. Comput. Appl. **10**(04), 32–34+39 (2020)
8. Silver, D., Huang, A., Maddison, C.J., et al.: Mastering the game of Go with deep neural networks and tree search. Nature **529**(7587), 484–489 (2016)
9. Silver, D., Schrittwieser, J., Simonyan, K., et al.: Mastering the game of go without human knowledge. Nature **550**(7676), 354 (2017)
10. Data source website: http://www.fierz.ch/download.php, Last accessed on 2014/12/10
11. Kumar, C., Fogel, D.B.: Evolving neural networks to play checkers without relying on expert knowledge. IEEE Trans. Neural Netw. **10**(6) (1999)

# Chapter 17
# Research on Intelligent Health Diagnosis of Tunnel Structure System Based on Wavelet Neural Network

**Ruijun Li and Yaqing Shi**

**Abstract** This paper proposes to establish a model of wavelet neural network to diagnose and predict the health of the tunnel structure. First, the dynamic response signal of the tunnel structure can be collected by wavelet transform, so the feature that best reflects the defect structure can be extracted. The node coefficient of wavelet packet can be used as the feature input vector of the neural network structure. Establish a wavelet neural network model, adjust the MSE value of the model by adjusting the relevant network and setting parameters, get the optimal model, and finally, use the trained neural network model to simulate effectively. The results show that this method has a good effect on the health diagnosis of the tunnel structure, and it can be used in subsequent engineering practice to be popularized.

## 17.1 Introduction

With the rapid economic development, the infrastructure construction is playing an increasingly important supporting role, and the high-speed railway has become a business card of China. Among them, the construction of tunnels has become indispensable, and the difficulties and challenges encountered in the construction process are also increasing. As we all know, most tunnels are composed of rock mass, and the health diagnosis of the tunnel structure is actually mainly the diagnosis of the quality of the surrounding rocks, so how to accurately diagnose the health of the tunnel structure has become our research problem and the direction of our efforts. The wavelet neural network, which combines wavelet analysis technology with artificial neural network, will undoubtedly have broad application prospects in the field of

R. Li (✉)
Hebei University of Architecture, Zhangjiakou 075031, China
e-mail: Liruijun110@163.com

Zhangjiakou Geotechnical Engineering Technology Innovation Center, Zhangjiakou 075000, China

Y. Shi
Zhangjiakou University, Zhangjiakou 075000, China



L. C. Jain et al. (eds.), *Smart Communications, Intelligent Algorithms and Interactive Methods*, Smart Innovation, Systems and Technologies 257,
https://doi.org/10.1007/978-981-16-5164-9_17

**Fig. 17.1** The equipment graph of IES

intelligent diagnosis of tunnel structural defects. Based on rock mass engineering, the model of wavelet neural network can be established to conduct neural network training and make accurate prediction for the tunnel structure [1, 2].

## 17.2 Experimental Process

### *17.2.1 Test Equipment and Basic Principles*

This project uses the equipment of impact echo scanner (IES), as shown in Fig. 17.1. IES adopts an original rolling sensor or automatic impact, which only needs to be tested while walking at a slower walking speed. It can scan and test the internal defects of various structures at a nearly continuous speed [3].

### *17.2.2 Specimen Production and Testing*

Three identical cylindrical rock cores were selected as test specimens on site, with a diameter of 50 mm and a height of 100 mm. Among them, specimen 1 and specimen 3 are intact rock samples, and specimen 2 is a rock sample that artificially simulates defects. Specimen 1 is the best specimen carefully selected by testers from a pile of intact specimens after undergoing mechanical testing and careful appearance observation. Specimen 2 is artificially penetrated in its longitudinal length direction by testers, with a crack width of 2 mm and a depth of 25 mm to simulate a rock sample with cracks. Specimen 3 is a simulated specimen carefully selected by testers from the pile of samples containing specimen 1. Afterwards, the acceleration sensors were arranged on the test points of the three specimens, and the excitation force was applied to the surface of the three specimens, which was chosen as a half-sine wave. The stress wave generated by the exciting force would propagate along the inside of the test piece. The propagation and reflection signal of the stress wave will be detected by the acceleration sensor attached to it, and the reflection information would be collected after receiving, amplifying and data-processing. Each test piece needs to be tested repeatedly with the same method until the waveform of each test piece is roughly the same.

## 17.3 The Establishment of Wavelet Neural Network Structure

The wavelet neural network consists of 3 layers as follows. (1) The input layer that receives the diagnostic signal data of the tunnel structure. A vector can be used to express the tunnel structure with a large amount of diagnostic information. This vector is composed of the node wavelet packet coefficients, which can be extracted by the wavelet packet decomposition of the reflected wave signal of the tunnel structure. (2) The hidden layer, which is alike to a "black box" operation to process the data. (3) The output-layer, which can be used to judge the integrity of the tunnel structure. The vector [1,0] can be revealed the integrity in this paper. For example, 1 means the intact tunnel structure, and 0 means the defective one. The integrity of the detected tunnel structure is judged according to the training result and the correlation between 1 and 0 [4].

## 17.4 Feature Vector Extraction of Wavelet Neural Network

EI and ED can be used to, respectively, represent the energy of each node of the wavelet packet of the intact specimen 1and the defective specimen 2. EC is also used to represent the magnitude change of the sub-signal energy of the $j$th ($j = 0, 1, \ldots, 31$) node of the fifth layer, that is EC = (EI − ED)/EI, which can indicate the measure whether the sub-signal energy of the jth node would have increased or decreased. The bar graph can be drawn by MATLAB programming shown in Fig. 17.2 for easier identification. From Fig. 17.2, it is easy to see that the energy transformation of the ninth node is the largest, so the wavelet packet node coefficient of this node should be regarded as the input feature vector of the neural network. The node wavelet packet coefficients of specimen 1 and specimen 2 can be, respectively, used as the input sample data for the intact tunnel structure and the defective tunnel structure, and the node wavelet packet coefficients specimen 3 are used as the simulation data.

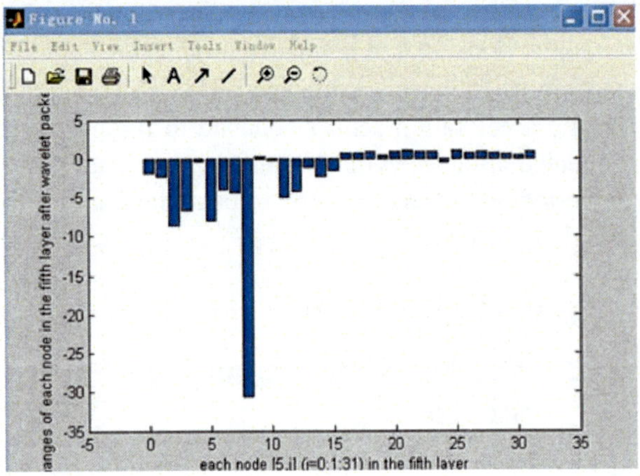


**Fig. 17.2** The bar graph of energy change of each node on the fifth layer

## 17.5 Training of Wavelet Neural Network

After repeated training and trial calculations, the neural network structure is selected, as shown in Fig. 17.3. Because the data in the input vector and the expected output data are too different in order of magnitude, the sample data and simulation data must be normalized. After repeated training and trial calculations, the number of neuron layers is selected as shown in Fig. 17.3. Select the number of neurons in the first layer as 32, and the transfer function as the LOGSIG function. Select 5 as the number of neurons in the second layer, and the LOGSIG function as the transfer function. Select 1 as the number of neurons in the third layer, and use PURELIN as the transfer function, as shown in Fig. 17.4.

The node wavelet packet coefficients of specimen 1 and specimen 2 are used as input data, and the expected value is [1,0], as shown in Fig. 17.5. Select the network type of feed-forward backprop and appropriate training parameters, as shown in the left of Fig. 17.4. The maximum of cycles is 50, and the minimum expected error is 1e−010, as shown in the right of Fig. 17.4. It can be seen from Fig. 17.5 that the error between the actual output value and the expected value is only 1.8887e−009, which is very near, indicating that the training result is ideal. Based on this, it can be inferred that the network structure and the training parameters are quite reasonable.

The node wavelet packet coefficients of the reflection signal of specimen 3 are used as simulation data, and the network simulation result is shown in Fig. 17.6. It can be seen from Fig. 17.6 that the output value is 0.8794, which is quite close to 1. Based on this correlation between 1and 0, it can be analyzed that the tunnel

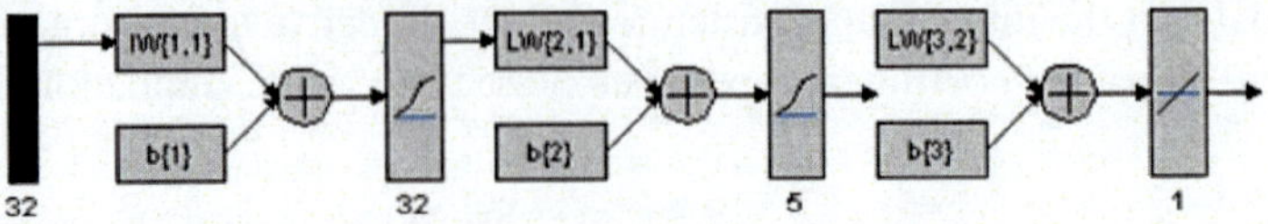


**Fig. 17.3** Neural network structure established diagram

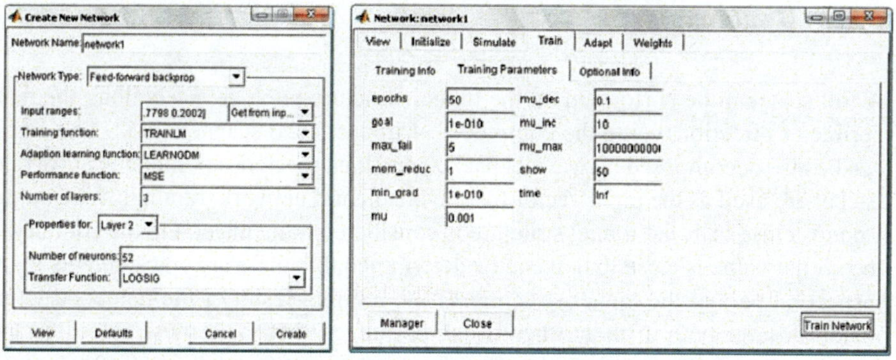


**Fig. 17.4** Training parameters of the wavelet neural network

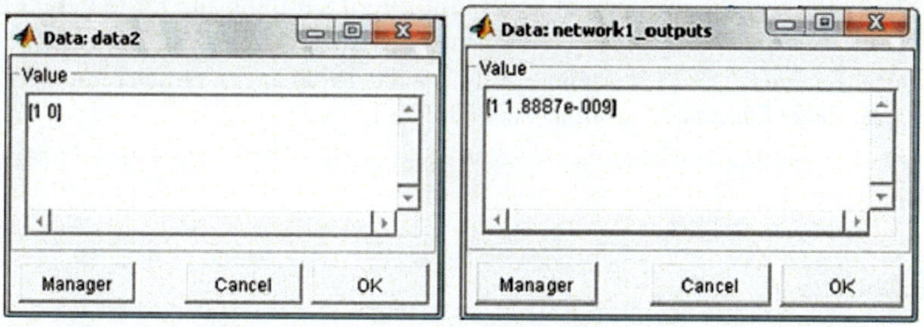


**Fig. 17.5** Comparison of actual output value and expected value

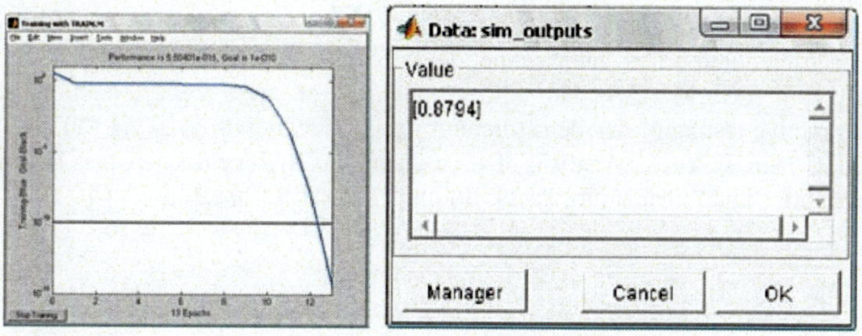


**Fig. 17.6** The simulation training results of the wavelet neural network

structure type of specimen 3 should be an intact tunnel structure, which is completely consistent with the actual results. This further proves the rationality and reliability of the established wavelet neural network, which can be promoted accordingly.

## 17.6 Conclusions

A full scan can be performed on the tunnel structure to be detected along the half-perimeter direction, using the equipment of impact echo scanner (IES). The signal can be also decomposed by wavelet packet, and then the node coefficients of wavelet packet are used as the input vector for wavelet neural network training. If the output value is close to 1, the tunnel structure is considered to be intact. On the contrary, if the output value is close to 0, it can be determined that there are crack defects in the inspection part of the tunnel structure. If the training result of the tunnel structure is not close to both of them, then it can be judged that there must be a defect in this tunnel structure, but the defect type cannot be identified yet, and further sample training and research should be needed. As a result, we can conclude that the use of wavelet neural network must realize the health diagnosis of tunnel structural defects, with the collecting of continuous accumulation of samples and large defect data.

**Acknowledgements** This work was financially supported by Zhangjiakou Science and Technology Bureau and Hebei Education Department (ZC2021224).

## References

1. Zeng, D.J., Liu, W., Du, Y.B.: Study on prediction model for deformation of unsymmetrical loading tunnel in bedding strata based on artificial neural network. Subgrade Eng. **3**, 170–174 (2020) [in Chinese]
2. Zhao, X.P., Sun, H.M., Zhang, S.X., Zhang, B.: Application of BP neural network in predicting ground settlement in urban shield tunneling. Subgrade Eng. **4**, 170–175 (2020) [in Chinese]
3. Li, R.J., Bai, R.S., Ma, J.S., Hu, J.L.: Application of stress wave reflection method in the non-destructive testing of wooden structure. Archit. Technol. **8**, 752–754 (2015) [in Chinese]
4. Wang, X.C., Shi, F., Yu, L.: Analysis of 43 cases of MATLAB neural network. Beijing University of Aeronautics and Astronautics Press, Beijing (2019) [in Chinese]

# Chapter 18
# Data on Cloud Practice Based on Alibaba Cloud IOT Platform

**Lei Wang**

**Abstract** Human beings are constantly carrying out creative thinking, constantly developing new and interesting things, changing the pattern of the world. Such a world is wonderful. As an emerging technology, the Internet of things is also adhering to this concept in the continuous development. In the whole Internet of things gateway technology, the Internet of things platform technology has become more and more important, and its application is also more and more widely pan, this article will borrow the introduction of Alibaba cloud's Internet of things platform, hoping that everyone can open up their own thinking, innovative development of more innovative creative products. Internet of things platform is a set of tools and services that enable developers to develop and run applications. Internet of things (IOT) platform is the middleware layer, which is responsible for obtaining data from sensors and terminal devices, generating meaningful results and operations. It usually provides device software development kit (SDK) or application programming interface (API). Developers can connect to any hardware platform through the platform and use cloud-based services.

## 18.1 Introduction

The Internet of things refers to the Internet that connects all objects together, or IOT for short. Due to the increasing application of Internet of things, the importance of Internet of things voucher is becoming more and more prominent. Hundreds of platform as a service (PaaS) providers have emerged in the world. All kinds of traditional IT enterprises, communication operators, communication equipment providers, Internet enterprises, industrial solution providers, and new start-ups are swarming into this market. The three cloud computing service modes of infrastructure as a service (IaaS), platform as a service (PaaS), and software as a service (SaaS) are gradually completed through the improvement of functional system. The following three modes are mainly introduced [1].

---

L. Wang (✉)
Shanghai Jianqiao University, 1111 Hucheng Ring Road, Shanghai, China
e-mail: 03010@gench.edu.cn



L. C. Jain et al. (eds.), *Smart Communications, Intelligent Algorithms and Interactive Methods*, Smart Innovation, Systems and Technologies 257,
https://doi.org/10.1007/978-981-16-5164-9_18

IaaS provides computing infrastructure (server, network technology, storage, and data center space) as a service to customers, including providing operating system and virtualization technology to manage resources. Consumers can obtain services from perfect computer infrastructure through Internet.

PAAS refers to the platform for software development. As a service, suppliers provide services beyond infrastructure, and a complete set of solutions as software development and operation environment are submitted to users in SaaS mode. Therefore, PaaS is also an application of SaaS mode. However, the emergence of PaaS can accelerate the development of SaaS, especially the development speed of SaaS applications.

SaaS is a delivery model in which applications are hosted as a service and provided to users through the Internet to help customers better manage their IT projects and services, ensure the quality and performance of their IT applications, and monitor their online business.

## 18.2 Types of IOT Platforms

At present, Internet of things platforms can be roughly divided into four categories, namely, communication field, internet field, software system service field, and vertical field.

Alibaba cloud platform provides users with a large number of cloud computing applications, including cloud computing foundation, database, security, big data, artificial intelligence, Internet of things, development and operation and maintenance, enterprise applications, and industry engines. Users can choose the corresponding products according to their actual needs. In terms of Internet of things platform, Alibaba cloud provides enterprise Internet of things platform and life Internet of things platform (Feiyan). Both of them provide cloud service platforms provided by IOT for Alibaba cloud. The two platforms have their own advantages. The enterprise Internet of things platform provides atomized equipment access capability, which is suitable for users with strong cloud development ability and can be used in various industries. The IOT platform provides device access capability, and public version app and interface free of development. It is more suitable for consumer-level intelligent device developers with low development threshold. It can quickly realize the intelligence of consumer-level devices, such as smart home appliances, wearing, home furnishings, etc. Alicloud Feiyan platform is an intelligent single product development platform, which is suitable for rapid access and operation of single products. Users who log in with the same Alicloud account will automatically synchronize all products and devices created on the IOT platform. The products created in the IOT platform can also be transferred to the life IOT platform by manually switching the charging mode. Alibaba cloud Internet of things platform provides safe and reliable connection and communication capabilities for devices, connecting massive devices downward to support equipment data collection in the cloud, providing cloud API upward, and sending instructions to the device side by calling cloud API, so as to realize remote control [2].

## 18.3 Basic Functions of Alibaba Cloud Internet of Things Platform

IOT platform can provide device access, device management, security monitoring, rule engine, and other functions, enabling all kinds of IOT scenarios and industry developers. The IOT platform supports the connection of massive devices to the cloud, and the devices communicate with the cloud stably and reliably through the IOT hub. It provides complete equipment life cycle management functions, supporting device registration, function definition, data analysis, online debugging, remote configuration, firmware upgrade, remote maintenance, real-time monitoring, group management, device deletion, and other functions. It provides multiple protection to effectively ensure the security of equipment and cloud data, mainly including identity authentication and communication security authentication. The rule engine of IOT platform includes server-side subscription, cloud product circulation function, and scene linkage.

## 18.4 Basic Operation Steps of Alicloud Platform

### *18.4.1 Register as an Alicloud User*

First of all, you can register on the front page of Ali cloud. There are two ways of registering, one is using Alipay account to register quickly, and the other is using account password. After registration, you need to bind your mailbox and complete real-name authentication to better use the relevant functions of Alicloud platform.

### *18.4.2 Add Products and Devices*

When creating a product, it is required to input the product name, category, node type, networking method, data format, authentication method, etc., after the product is created, and you can add the device and define the object model for the product. After the creation of the device, the Alicloud platform will automatically generate a triple of the device, that is, the main content of the device certificate. The certificate content of the device should include product key, device name, and device secret can be viewed by users and copied by using the one click copy function. Triples are important credentials for data transmission between the device and the Alicloud platform, so they must be kept properly.

## 18.5 MQTT Software Connectivity Test Steps

MQTT refers to message queuing telemetry transmission, MQTT.fx. It is an MQTT client tool based on eclipse and written in Java language. It supports subscribing and publishing messages through topic. When setting the MQTT proxy profile, there are three important locations of content setting, which are access domain name, client Identity document (ID) number, and user credentials [3].

Content format of access domain name:

${YourProductKey}.iot-as-mqtt.${YourRegionId}.aliyuncs.com;

The client ID number uses the underlying protocol message of MQTT, and the specific format is as follows:

${clientId}|securemode=3,signmethod=hmacsha1|;

The user name and password should also be set in the user credentials, in which the user name, device name, symbol (&), and product key are composed. The format is as follows:

${YourDeviceName}&${YourProductKey}

The password is made by splicing and encrypting the parameter values, which can be generated by using special tool software. After setting the parameters, click connect to connect to the Alibaba cloud Internet of things platform, as shown in Fig. 18.1.

Downlink communication data test, sending messages from the Internet of things, and MQTT.fx receiving information in software, testing MQTT.fx is the connection between the software and the Internet of things platform successful. Uplink communication test, passed MQTT.fx the software sends messages to the Internet of things platform to test connectivity use MQTT.fx after the software realizes simple uplink and downlink communication of data, and you can try to use topic to report the virtual device properties. The specific operations are as follows: click publish, enter a topic

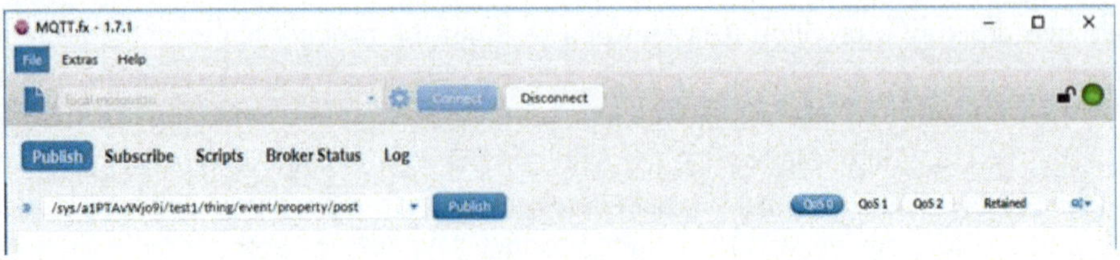


Fig. 18.1 Parameter configuration and connection

with publishing permission for the device, and enter the following code below, where "humidity": 60 means the simulated humidity is 60%, and "temperature": 28 means the simulated temperature is 28° C, click publish.

```
{
"params":{"Humidity":60,"Temperature":28},
"method":"thing.event.property.post".
}
```

## 18.6 Conclusion

The use of Internet of things cloud platform greatly simplifies the development cost and learning cost of users and can build the Internet of things platform in the shortest time with the least time and experience. Such changes can help users focus on creative thinking and take innovation as the main research direction. The implementation process is not important, and the novel thinking is the key. Alibaba cloud Internet of things. The platform only provides a tool, a means for users to achieve their goals. People's creativity and ideas are the key. Only when people shackle, the shackles of technology and let their thinking roam in the creative space, can they have better development. The Internet of things technology is originally proposed to transform human life, whether it is the Internet of things smart home, the Internet of things smart agriculture, or the Internet of things Internet intelligent logistics is to bring people a better life experience, so the application based on the Internet of things platform can reduce the threshold of technology so that more users can realize their own life value.

## References


1. Han, C.Y.: Development and design of internet of things teaching platform based on Alibaba cloud. Electron. Test. 86–88 (2020)
2. He, M.: Design of household electricity monitoring system based on Alibaba cloud. Mech. Eng. Autom. 135–137 (2020)
3. Cai, Z.W.: Design of smart home system based on mqtt. J. Yuncheng Univ. 29–32 (2020)

# Chapter 19
# Remote Intelligent Guidance System Based on Augmented Reality

**Kaijia Luo, Zhenxiang Li, Zhiqi Guo, Yanan Zhang, and Zhang Lu**

**Abstract** The special environment of the Space Station and complexity of space equipment require more efficient and accurate on-orbit maintenance methods. Space equipment failures can be solved by adding maintenance plans in the way of induced maintenance. However, the induced maintenance plan cannot be added unlimitedly, time expenditure and cost consumption should be considered. Using remote communication with experts on the ground can effectively save costs and improve maintenance efficiency. Under current conditions, astronauts cannot connect experts through conventional communication methods. To sum up the foregoing, designed remote intelligent guided maintenance system based on Augmented Reality, which is applied to the actual work field where remote experts guide astronauts to coordinate maintenance. Compare and choose the right software, hardware, and development platform. Realize remote data processing and transmission, world-earth simulation communication, remote virtual information addition, and information storage. The astronaut terminal and the remote expert terminal realize audio and video data transmission based on socket multi-threaded communication, which breaks through the limitations of astronaut's local operation. Develop and test the system on Unity platform, software runs stably, and the remote auxiliary maintenance tools are intuitive and convenient, which can effectively realize the remote collaboration between ground and the Spacc Station.

K. Luo (✉) · Z. Li · Z. Guo · Y. Zhang · Z. Lu
Key Laboratory of Spacc Utilization, Chinese Academy of Sciences, Beijing 100094, PR China
e-mail: luokaijia@csu.ac.cn

Technology and Engineering Center for Space Utilization, Chinese Academy of Sciences, Beijing 100094, PR China

University of Chinese Academy of Sciences, Beijing 100094, PR China

Z. Lu
e-mail: zhanglu@csu.ac.cn

 
L. C. Jain et al. (eds.), *Smart Communications, Intelligent Algorithms and Interactive Methods*, Smart Innovation, Systems and Technologies 257,
https://doi.org/10.1007/978-981-16-5164-9_19

## 19.1 Introduction

The Space Station is a large-scale spacecraft which can be used for long-term work and residence for astronauts. It is responsible for conducting near-Earth space science and technology experiments, comprehensive development and utilization of space resources, being special environment of vacuum, high radiation, and weightlessness. It has the characteristics of few guarantee resources and high safety requirements. The space mission is complex and changeable. In order to ensure the long-term stable operation of the Space Station, it's necessary to adopt preventive maintenance and corrective maintenance strategies and rely on astronauts' in-orbit maintenance [1].

Improving maintenance methods is a necessary strategy to improve the reliability of space station equipment. Using Augmented Reality [2] technology can superimpose virtual images in the real environment and achieve information enhancement. It can optimize the complex maintenance operation process, reduce mis-operation rate, and improve maintenance efficiency [3]. At present, the induction maintenance system based on Augmented Reality mainly studies individual induction methods which need making and reserving related virtual information in advance, mostly used in maintenance personal training [4]. Sudden failures and uncertainties determine astronauts' knowledge reserve and induced solution plan definitely have certain limitations in space on-orbit maintenance. On this basis, extending training time and increasing maintenance guidance plans will increase costs and time loss.

Turn to experts with rich relevant experience can assist system to further realize intelligent induction. The space station's application environment is not sufficient to be staffed with multiple technical experts and therefore cannot be used for local help. Combined with remote guidance technology, when astronauts face unexpected problems, asking remote experts for help is a good solution.

Common ground remote collaboration ways include text or image sharing, voice calls, and video conferencing [5]: Remote interaction methods based on text or images lack real-time performance, voice calls are difficult to eliminate ambiguity. Video conferencing based on voice and real-time video sharing has become an ideal remote interaction method [6]. Remote collaboration between astronauts on the Space Station and remote experts on the ground can realize by learning from the ground remote collaboration. It only needs to solve the remote communication problem between space and earth links.

Real-time sharing visual information during remote collaboration can help enhance mutual understanding between astronauts and technical experts, improve work efficiency [7]. However, remote experts can only get astronaut's perspective picture and information, cannot fully grasp the real-time situation, only observe the picture for voice guidance, and lack proper interaction. Add remote annotations can achieve clearer remote guidance, such as arrows and other induced maintenance information.

In summary, this article is facing the remote maintenance guidance needs of the Space Station, based on existing research, draw on remote collaboration technology under similar ground scenes. Combine guidance system based on Augmented

Reality with remote collaboration, propose and design a Remote Intelligent Guidance Maintenance System for the Space Station.

## 19.2 System Design

The Remote Intelligent Guidance Interactive System combines Augmented Reality and remote guidance to assist astronauts in completing space on-orbit maintenance tasks.

### *19.2.1 System Framework*

The Remote Intelligent Guidance Interactive System is composed of three important parts (see Fig. 19.1): astronaut-side software, server-side software, and experts-side software.

Astronaut side software running on the Space Station stores pre-store troubleshooting plan for induced maintenance. During the orbit maintenance process, astronauts wear head-mounted display (HDM) to perceive surrounding scenes and status changes of maintenance objects through natural interactive means and inquire whether there is a corresponding troubleshooting plan. If current failure can be solved individually, the corresponding maintenance data will be stored on the local server equipped in the cabin. If cannot solve, astronaut side software will collect local audio and video data, forward data through server, send captured real-time audio

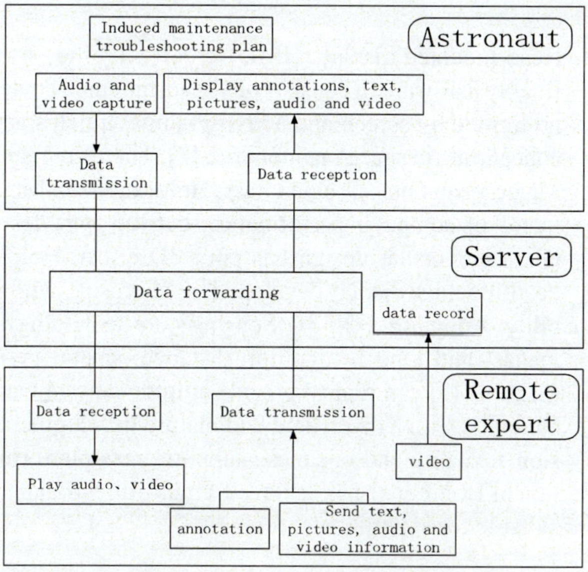


**Fig. 19.1** System framework

and video streams, fault information, and three-dimensional spatial data to online remote experts for helping. Receive and display all kinds of auxiliary maintenance information sent by remote experts in real time.

The expert terminal software keeps real-time monitoring, waiting for connecting from astronauts. After receiving the communication request, server receives real-time audio and video data streams, as well as additional information such as space positioning. Experts use local resources and their own personal knowledge to provide real-time maintenance guidance. Guidance methods include text input, picture transmission, voice dialog, etc., and can further add certain digital maintenance guidance information to the real-time video stream and send it to the astronaut via the server.

The server is deployed on the ground, waiting for the real-time audio and video data stream of the astronaut to go down, and transmitting through the space-earth communication link, sending the data to the expert workstation on the ground. Wait for remote experts to add annotations, send text, pictures, and other data, and transmit the processed data to the astronaut. The server also stores and manages all necessary maintenance data information.

The Remote Intelligent Guidance Interactive System based on Augmented Reality gives full play to the respective advantages of astronauts and remote experts: astronauts have maintenance conditions and a complete understanding of the working space; remote experts have sufficient knowledge and rich experience to assist astronauts in completing maintenance tasks and can combine the abundant relevant reserve resources on the ground. The development of a remote intelligent induction interactive system can improve the efficiency of space on-orbit maintenance, accurately and quickly solve sudden failure.

### *19.2.2 Hardware and Platform Selection*

Head-mounted display (HMD) is currently the fastest-growing type of Augmented Reality hardware. It consists of a head-mounted device and a matching micro-display, no limited by screen and activity radius. It helps astronauts to free their hands and concentrate in space maintenance [8]. There are many AR hardware devices, such as Magic Leap One, Google Glass, HoloLens, etc. Magic Leap One meets the requirements of computer performance, but the split design is difficult to carry. Google Glass monocular design has poor 3D effect. HoloLens is an integrated binocular head-mounted display developed by Microsoft, with a stable structure and easy portability. Adopting TOF 3D Sensing core technology, good anti-interference performance, and long recognition distance; spatial perception positioning technology (SLAM) [9] can realize precise attitude determination and position determination; equipped with a diversified and mature development environment, with high expansion sexuality and openness support cross-platform cooperation. The new generation of HoloLens2 has achieved a multi-directional technical upgrade, which is more suitable for experimental research.

The development platform for Augmented Reality content is mainly based on three engine tools: Unreal 4 from Epic Games; CryEngine V from Crytek; and Unity from Unity Technologies. Unreal Engine is compatible with most platforms but bulky, C++ language development is not easy to learn. Cry Engine development is free but has low efficiency and poor reliability. Amazon Lumberyard cannot support mobile platform development. The Unity engine has low requirements for hardware equipment, use C# programming, easy to get started, simple to operate, and free to develop; strong community support and third-party plug-ins, compatible with all platforms.

Compared with other development platforms, Unity's support for HoloLens is more mature, related materials and SDKs are more complete. Tools such as Mixed Reality Toolkit (MRTK) allow developers to easily deploy projects to HoloLens through Universal Windows Platform (UWP). In summary, Unity is the best platform to develop a remote intelligent induction interactive system based on Augmented Reality.

## 19.3 Software Overall Design

### *19.3.1 Software Design for Astronaut Part*

The astronaut side software uses Visual Studio 2019 as the development environment, C# language development, and use socket communication based on UDP communicate with server. The program design mainly includes data acquisition, data processing, and remote sending modules. Realize the functions of data collection, sending, and reading on the astronaut terminal.

On the display interface of astronaut, set server's IP address and port, create a data stream UDP socket; Sendto() sends data packets to the server's IP and port, the real-time audio and video data is cut and sent; Recvfrom() waits for the server process to return the processing result, receive remote data from remote expert; after communication ends, call Close to close the socket (see Fig. 19.2a). Since the UDP protocol is used to transmit audio and video streams, there is no need to maintain the connection before sent, destination address and port will be carried with data, which reduces resources occupied by maintaining the network connection. The data delivered to the UDP protocol stack will be sent out as soon as possible to ensure transmission efficiency. Uncompressed digital video has a huge amount of data, difficult to store, and difficult to transmit. The data can be compressed by removing redundant information. Protobuf can compress and decompress audio and video signals.

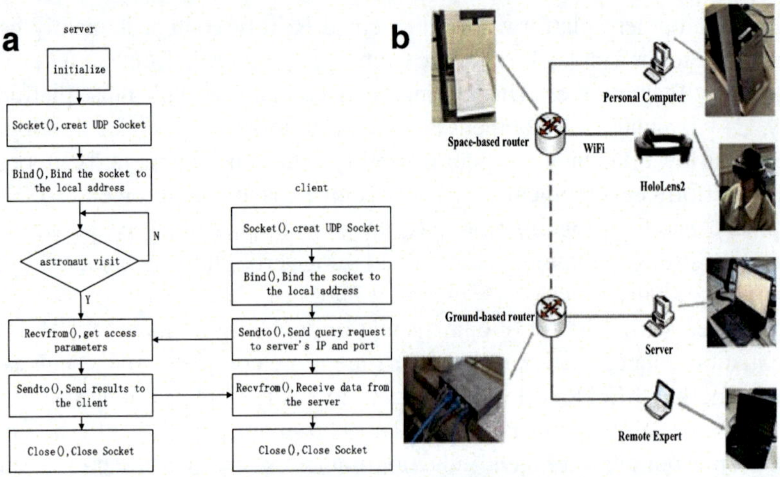


**Fig. 19.2** **a** socket transmission process. **b** Ground simulation device

### *19.3.2 Software Design for Remote Expert Part*

The remote expert software also uses Visual Studio 2019 as the development environment and C# language development, which mainly includes: data stream receiving and sending, remote information design and addition, file storage, and other functional modules. Realize the functions of receiving data and information transmitted by the astronaut, sharing the remote screen of the astronaut in real time, calling the local knowledge base for production and remote release of induced maintenance information, and storage of maintenance information.

In the main display interface of astronaut side, set IP address and port of server, call socket to create a datagram socket; call blind to bind it with server address; wait for astronaut side process to send a request through recvfrom(), and then remote expert side In an infinite loop state; after receiving the datagram sent by astronaut process, process it, call sendto() to return the processing result to astronaut process, and return to blind state. After process ends, call close to cancel the socket and end communication.

After remote expert completes data reception and transmission, it actively calls the OpenFileDialog class to start the file dialog, opens the default storage path and storage format through Initial Directory and Filter, waits for expert to input file name, calls File Stream to store stream which can read it again.

### *19.3.3 Design and Simulation of Space-Earth Link*

The space-earth integrated space Internet is the technical basis for real-time communication of remote intelligent induction interactive systems. A local area network based on the Internet protocol is formed in space station, and Transmission Control Protocol (TCP-IP) is used for ground data transmission. The space-based system and ground-based system are connected via Ethernet. Figure 19.2b shows a schematic diagram of transmission of space and earth information between astronauts and remote experts in the Space station.

The astronauts are equipped with induction maintenance terminal HoloLens2 and computers equipped with the space station are connected via Wi-Fi via Ethernet as the astronaut terminal; ground expert system sets up a server and multiple computers as the expert terminal. The astronaut terminal and the expert terminal transmit data through the space-to-earth communication link. Use the router to connect server, PC computer, HoloLens2 to build a ground LAN for simulation, as shown in the figure. Storage of audio and video information requires a database, and users can perform operations such as adding, querying, updating, and deleting data in the file. Storing and managing large amounts of data requires Relational Database Management System. The server database uses MySQL to store data in different tables, which increases speed and flexibility. SQLite software library can be used for local data, which is self-sufficient, serverless, and zero configuration.

## 19.4 Results and Discussion

After software is running, enter login interface, enter IP address and port number, log in to astronaut software. Query and connect with online experts (a), establish a video communication connection with remote experts (b), and share expert video images in real time. The remote expert logs in to software on PC and waits for remote invitation. After being connected, the voice and video sharing (c) can be realized, and remote marking information can be sent to guide (d). After communication is ended, wait for the next connection. And information is stored in a local folder (Fig. 19.3).

## 19.5 Conclusions

This paper constructs a remote expert intelligent guidance interactive system based on Augmented Reality technology and remote collaboration. Socket network programming technology is used to construct a network transmission module, and Protobuf performs data encoding and decoding, Vuforia and MRTK realize remote labeling, set MySQL and SQLite servers, and realize ground simulation of the remote communication architecture based on Ethernet. The system has been verified in real-time through remote communication experiments, the video refresh rate is high, and basic remote virtual auxiliary information can be added. The remote intelligent interaction system can effectively realize the synchronous or asynchronous remote interaction between astronauts and remote experts. Solve both hands and establish appropriate

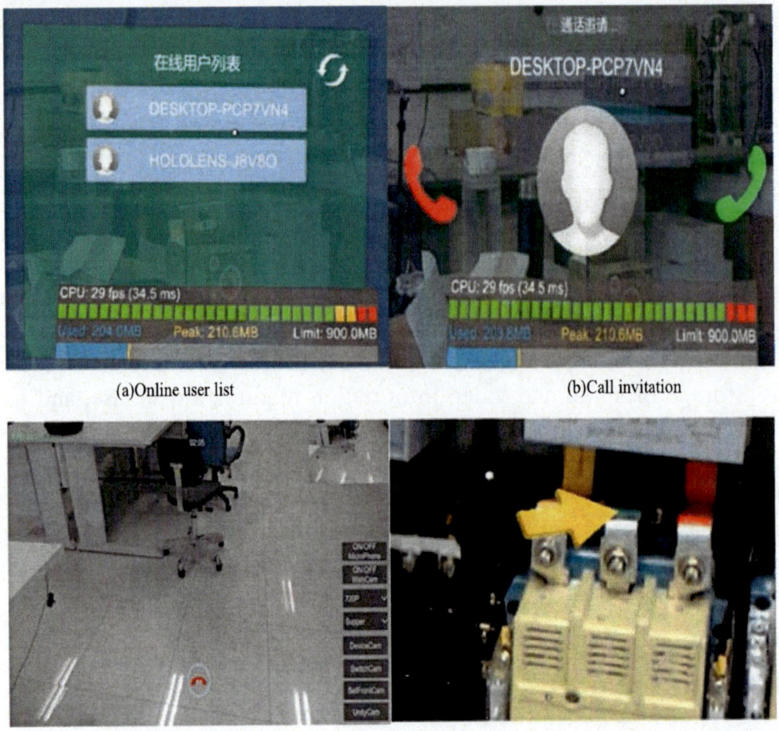


(a)Online user list (b)Call invitation

(c)Share video screen (d)Add remote annotation

**Fig. 19.3** An instance of a software application

communication channels, support remote maintenance and repair operations based on Augmented Reality, and improve resource utilization and maintenance efficiency.

## References

1. Ge, X.Y., Zhou, Q.X., Liu, Z.Q.: Assessment of space station on-orbit maintenance task complexity. Reliab. Eng. Syst. Safety, 193 (2020)
2. Azuma, R., Baillot, Y., Behringer, R.: Recent advances in augmented reality. IEEE Comput. Graphics Appl. **21**(6), 34–47 (2001)
3. Schwald, B., Starmate: Using augmented reality technology for computer guided maintenance of complex mechanical elements. Proceedings of EBusiness and EWork Conference, pp. 17–19 (2001)
4. Henderson, S.J., Feiner, S.: Evaluating the benefits of augmented reality for task localization in maintenance of an armored personnel carrier turret. In: Proceedings of the 2009 8th IEEE International Symposium on Mixed and Augmented Reality, pp. 19–22 (2009)
5. Karim, R., Candell, O., Söerholm, P.: E-maintenance and information logistics: Aspects of content format. J. Qual. Maintenance Eng. 15 (2009)
6. Mourtzis, D., Zogopoulos, V., Vlachou, E.: Augmented reality application to support remote maintenance as a service in the robotics industry. Procedia CIRP, pp. 46–51 (2017)

7. Zhong, X., Liu, P., Georganas, N.: Designing a vision-based collaborative augmented reality application for industrial training. Inf. Technol. 7–19 (2003)
8. Kim, H.Y.: The current status and development direction of mixed reality content. Cartoon Anim. Stud. 181–206 (2017)
9. Liu, Y., Dong, H., Zhang, L.: Technical evaluation of HoloLens for multimedia: a first look. IEEE MultiMedia **25**(4), 8–18 (2018)

# Chapter 20
# Research on Air Engine Air Circuit Fault Based on Convolutional Neural Network Feature Extraction

**Yi Ren Zhou, Zhi Qiang Guo, and Chao Zhou**

**Abstract** The research of aeroengine gas path fault diagnosis algorithm based on the combination of CNN-SVM is to convert the aeroengine gas path status data into a time-parameter matrix; then on the basis of a large number of normal samples, CNN is used to establish the engine gas path status feature extraction model; the features mined after multiple convolution and pooling operations are no longer directly connected to the classifier through the fully connected layer for classification, but instead use it as the state characteristics of the fault sample, and use SVM to classify these features. The detection rate of each type of fault is more than 90%. The CNN-SVM diagnostic method has broad application prospects in the engineering practice of aeroengine gas path fault diagnosis.

## 20.1 Introduction

The aeroengine gas path status data is a typical multi-dimensional time series data. When performing fault diagnosis on the aeroengine gas path, it is necessary to synthesize the change trends of various monitoring parameters in order to obtain more accurate fault diagnosis. At present, most fault diagnosis methods can only input vectors. The usual practice is to vectorize multi-dimensional time series data, and then use the corresponding algorithm for fault diagnosis, thus ignoring the correlation between parameters and the influence of parameter sequence. In 1989, Cun et al. Proposed the concept of convolutional neural network (CNN) [1]. The convolution operation in CNN can be used to directly input the matrix. For the engine gas path parameters, both in addition to the relationship between input parameters, the relationship of input parameters with time is also considered. CNN is a typical supervised deep

Y. R. Zhou · Z. Q. Guo (✉)
Nanchang Institute of Technology, Nachang 330044, China

C. Zhou
Jiangxi Province Energy Conservation and Environmental Protection Investment Company, Nanchang, China

 
L. C. Jain et al. (eds.), *Smart Communications, Intelligent Algorithms and Interactive Methods*, Smart Innovation, Systems and Technologies 257,
https://doi.org/10.1007/978-981-16-5164-9_20

neural network. At present, the diagnosis algorithms based on CNN are all end-to-end algorithms. The premise of these algorithms is that there are enough samples in the category, otherwise the network will be seriously over-fitted. The end-to-end diagnostic algorithm is not suitable for the diagnosis of aeroengine gas circuit faults.

In response to the above problems, this paper designs a fault diagnosis method of aeroengine gas path based on CNN and support vector machines (SVM). This method is based on a large number of actual health status data of the engine, using CNN to establish an engine gas path state feature extraction model based on parameter correlation and sequence, using this model to perform feature mining on engine gas path fault data, and as the input of SVM, so as to achieve the purpose of fault diagnosis. The model uses convolutional neural networks to consider the relationship between input parameters and the impact of time series and uses SVM to classify well to solve the problem of insufficient engine fault samples.

## 20.2 CNN and SVM-Based Aeroengine Fault Diagnosis Method

According to the characteristics of CNN's typical supervised deep neural network, the current CNN-based fault diagnosis models are directly using CNN for classification. The premise of this algorithm is that there are enough samples in each category, and the current convolutional neural network in other words, even if there are hundreds of samples in each category, the network will still fall into overfitting [2]. However, there are relatively few failure cases during the flight of an aeroengine, so it is not realistic to directly use CNN algorithm to classify the aeroengine gas path failure. To this end, this paper designs a fault diagnosis method of aeroengine gas circuit based on CNN and SVM. This method first uses CNN to establish an engine gas path state feature extraction model based on a large amount of actual health data and then uses this model to perform feature mining on engine gas path faults. The features extracted through multi-layer convolution and pooling operations are not then directly classify through the fully connected layer connection classifier, but use it as the input of the support vector machine for classification, so as to achieve the purpose of fault diagnosis.

### *20.2.1 Fault Diagnosis Process and Implementation Steps*

The fault diagnosis method based on CNN and SVM proposed in this paper is used to diagnose the aeroengine gas path faults (Fig. 20.1). The specific process is shown in Fig. 20.2. For convenience, the above method is referred to as CNN-SVM fault diagnosis method. The specific implementation process of engine gas path fault diagnosis using CNN-SVM method is as follows:

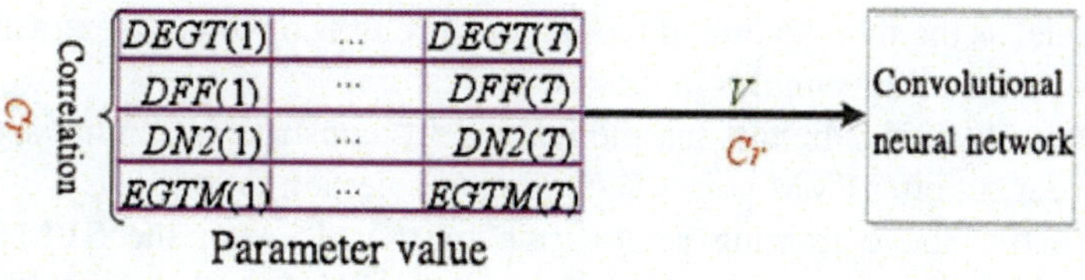


**Fig. 20.1** Schematic diagram of aeroengine state parameter matrix

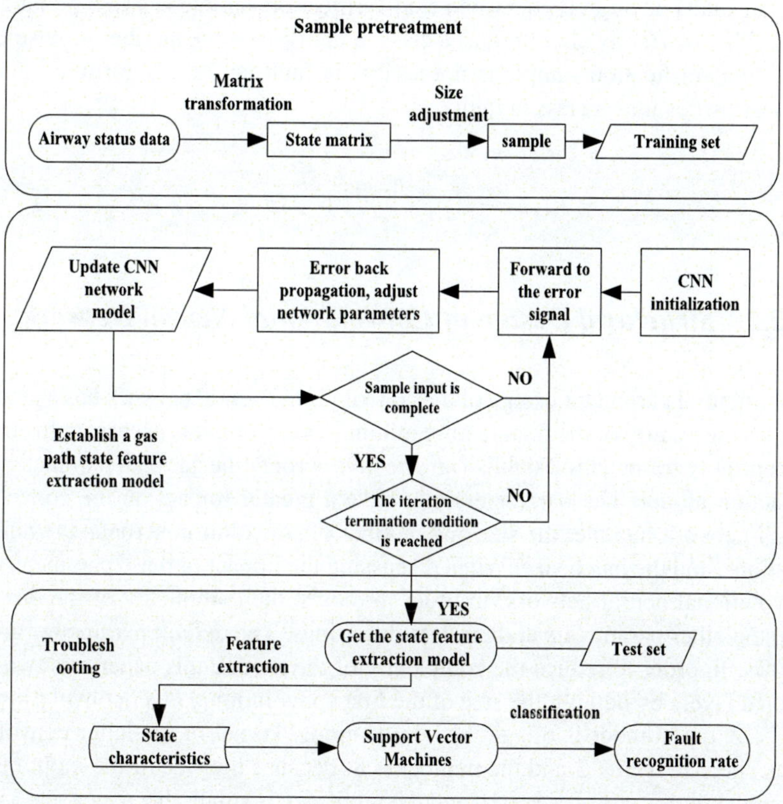


**Fig. 20.2** CNN and SVM-based aeroengine gas path fault diagnosis process

(a) Construct the state matrix of engine gas path data as shown in Fig. 20.1;
(b) Construct a training sample set and a test sample set, where the training sample set is all composed of the engine normal state data matrix, and its label is "0"; the test set is composed of the engine failure indication data matrix and the normal data matrix, the corresponding fault type shall be labeled accordingly;
(c) Use the training sample set to train the CNN network. After the network training is completed, keep the network parameters unchanged;
(d) Use the trained network to perform feature mining on the samples in the test set, and use the input vector of the first fully connected layer in the network

model as the final feature of the sample through convolution extraction, so that the status of all samples in the test set feature;

(e) After obtaining the test sample feature set, construct a training set and a test set for training SVM points according to a certain ratio [3];

(f) Use the above training set to train the SVM. After the SVM training is completed, use the above test set to test the SVM. In this paper, the accuracy value prec is used as an indicator to evaluate the effect of classification. The specific calculation formula is shown in formula (20.1): where tp is the number of faults accurately identified (the verification sample is a ia type fault, algorithm Correctly classify it as a class ia fault), fp is the number of false outliers (the verification sample is not a class ia fault, and the algorithm incorrectly classifies it as a class ia fault).

$$\text{prec} = \frac{\text{tp}}{\text{tp} + \text{fp}} \tag{20.1}$$

### 20.2.2 *Structural Design of Convolutional Neural Network*

The structural parameter design of the convolutional neural network has a great influence on the quality of the final feature mining. Therefore, it is necessary to choose an appropriate parameter to establish an effective aeroengine gas path parameter feature extraction model. The parameters that have a greater impact on the convolutional neural network include: the size and number of convolution kernels, the number of iterations, and the batch size. When discussing the impact of these parameters on the convolutional neural network, in order to reduce the complexity of the discussion, keep the other parameters at their default settings. The default parameters are set as follows: In order to reduce the complexity of the model, this paper sets two convolutional layers by default, the size of the first convolutional layer convolution kernel is $2 \times 2$, the number is 6, and the convolutional kernel of the latter convolutional layer The size is $3 \times 3$ and the number is 8; because the size of the input matrix of the model in this paper is $4 \times 10$ and the input size is small, this paper sets a pooling layer by default, the size of the pooling layer is $2 \times 2$, and the pooling method For average pooling, the pooling layer is connected to the latter convolutional layer; the batch size is set to 100 by default; the number of iterations is set to 100 by default. The following will analyze and study the quality of these parameters on the final feature extraction of the model to select reasonable parameters.

#### 20.2.2.1 Batch Size Design

When training a convolutional neural network, the loss function required to update the network parameters each time is not obtained from the full sample set training, but a random set of samples is randomly selected from the full sample set for training.

Such a set of samples contains the number is a batch size. If the batch size is too large, the number of iterations required to complete the training of a full sample set is reduced, but the convergence accuracy of the network will fall into different local extreme values; if the batch size is too small, the algorithm will run the risk of non-convergence, and once training The full sample set takes longer. Increasing the batch size within a reasonable range not only reduces the time required to train a full sample set but also accelerates the processing speed for the same amount of data, and as the batch size increases, the more accurate the network convergence direction is and the smaller the training oscillation.

This experiment mainly discusses the effect of batch size on CNN feature extraction ability. In order to ensure the comparison under the same standard, other parameters except batch size remain the same by default. In order to fully study the effect of batch size on CNN's ability to extract features, this paper sets the batch size to 5, 10, 20, 30, 40, 50, 60, 70, 80, 90, 100, 110, 120, 130, 140, 150, and 160. In order to eliminate the randomness of the experimental results, each experiment was repeated 5 times and the average of the experimental results was taken. The final result is shown in Fig. 2.14, when the bitch size is 5, 10, 20, 30, the algorithm does not converge. As can be seen from Fig. 20.3, when the batch size is between 50 and 100, the change in batch size has little effect on the accuracy of the final classification; when the batch size is greater than 100, as the batch size increases, Although EGT indicated that the failure recognition rate did not change much, TAT indicated the failure recognition rate, the normal sample recognition rate and the overall recognition rate were gradually reduced, but the HPT blade ablation failure recognition rate was significantly reduced. When the batch size is equal to 100, the TAT indicates the fault identification rate, the HPT blade ablation fault identification rate, the normal sample identification rate and the overall classification correct rate are all optimal,

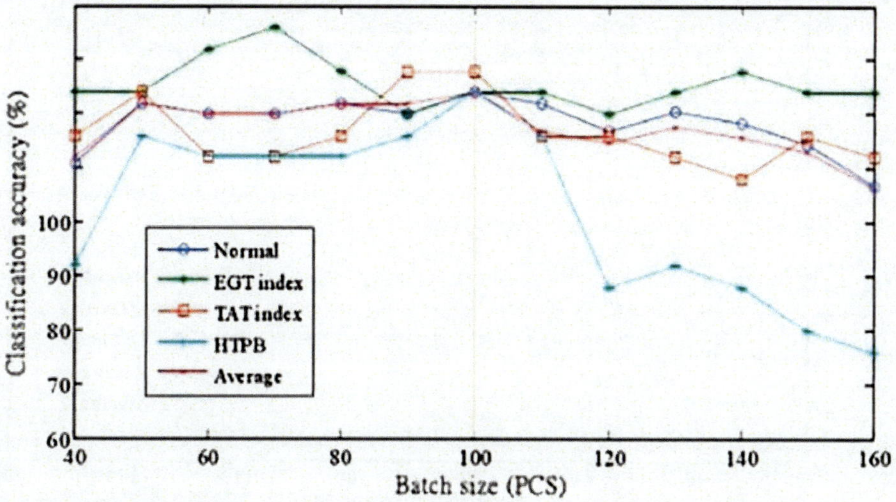


**Fig. 20.3** Relationship between batch size and classification accuracy

and the EGT indicates the fault identification rate also reaches more than 90%. Therefore, for the comprehensive fault classification accuracy and training time cost, the batch size can be set to 100 [4].

#### 20.2.2.2 Iteration Design

The neural network iterates through fitting and approximating samples continuously. If the number of iterations is too small, the fitting effect will be poor; if the number of iterations is too large, the network error will no longer decrease and the training time will continue to increase. Therefore, choosing the appropriate number of iterations, while meeting the diagnostic accuracy, also needs to reduce the training time. This experiment mainly discusses the effect of the number of iterations on CNN. In order to ensure the comparison under the same standard, except for the number of iterations, the batch size is set to 100 according to the above conclusions, and other parameters remain unchanged by default. In order to fully study the impact of the number of iterations on CNN's ability to extract features, this article sets the number of iterations to 1, 5, 10, 20, 30, 40, 50, 60, 70, 80, 90, 100, 110, 120, 130, 140, 150, and 160. In order to eliminate the randomness of the experimental results, each experiment is repeated 5 times, and the average value of the experimental results is taken. The experimental results are shown in Fig. 20.4. It can be seen from Fig. 20.4 that as the number of iterations increases, the correct rate of fault classification increases, especially the HPT blade ablation fault recognition rate increases significantly with the number of iterations. When the number of iterations is greater than 100, as the number of iterations increases, the failure recognition rate exceeds 90% and tends to

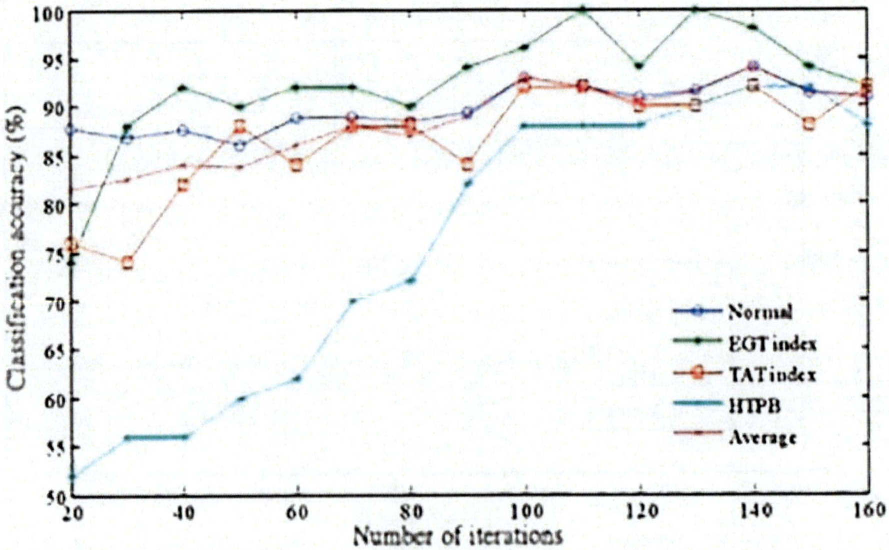


**Fig. 20.4** Relationship between the number of iterations and classification accuracy

**Table 20.1** Relationship between classification accuracy and size of convolution kernel

| Serial number | ($k_1$, $k_2$) | prec (EGTindex) | prec (TATindex) | prec (HPT_Blade) | prec (Normal) | Diagnosis correct number/Total number of samples |
|---|---|---|---|---|---|---|
| 1 | (2, 2) | 0.92 | 0.82 | 0.52 | 0.855 | 0.832 |
| 2 | (3, 2) | 0.94 | 0.82 | 0.92 | 0.87 | 0.886 3 |
| 3 | (2, 3) | 0.96 | 0.92 | 0.94 | 0.92 | 0.922 |
| 4 | (3, 3) | 0.96 | 0.8 | 0.9 | 0.9 | 0.9 |

be stable; when the number of iterations is equal to 140, the failure recognition rate reaches the optimal. Therefore, considering the training time and fault recognition rate comprehensively, under the sample size of this paper, it is most reasonable to choose 140 iterations (Table 20.1).

Design of the number of convolution kernels

The design of the number of convolution kernels corresponds to the number of feature maps. With several convolution kernels, several feature maps can be extracted. The number of convolution kernels is too small to fully mine the features of the input picture; the number of convolution kernels is too large. As the training parameters increase, the more complex the model, the number of required samples increases, and the training time increases. Therefore, it is necessary to select an appropriate number of convolution kernels, which can not only fully tap the features of the samples, but also minimize the complexity of the model. This experiment mainly discusses the influence of the number of convolution kernels on the ability of CNN feature extraction and selects the appropriate number of convolution kernels. When discussing the impact of the number of convolution kernels, in addition to the number of convolution kernels, the number of iterations should be set to 140, the batch size to 100, and the remaining parameters remain unchanged by default. Since two convolutional layers are designed by default, in general, the number of convolution kernels of the second convolutional layer is larger than that of the first convolutional layer. In addition, this paper finally uses SVM to classify the features, and the extracted feature dimension should not be too high, so the number of convolution kernels of the second convolutional layer is set to not exceed 10 at most. In theory, as long as the above two conditions are met, the number of convolution kernels of the two convolutional layers can be any combination, but this combination is too complicated, so the number of convolution kernels of the second convolutional layer is set during the experiment. It is 10, 9, and 8. In order to eliminate the randomness of the experimental results, each experiment is repeated 5 times, and the average value of the experimental results is taken. The specific experimental results are shown in Fig. 20.5. It can be seen from Fig. 2.16a–c that the classification accuracy increases with the number of convolution kernels of the two convolutional layers. When the number of convolution kernels in the second layer is fixed, as the number of convolution kernels in the first layer

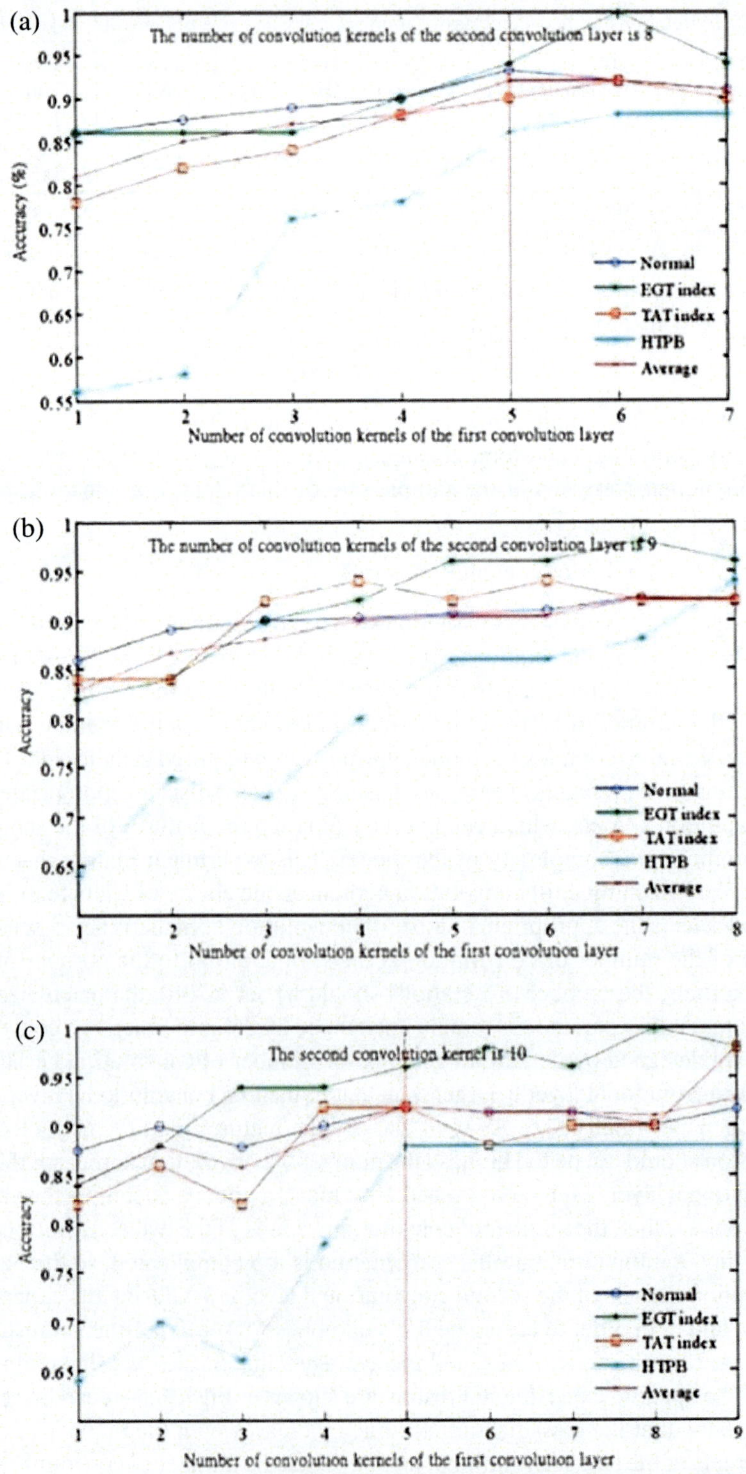


**Fig. 20.5** Relationship between classification accuracy and the number of two-layer convolution kernels. **a** The second convolution layer has 8 convolution kernels. **b** The number of convolution kernels of the second convolutional layer is 9. **c** The second convolution layer has 10 convolution kernels

increases, the classification accuracy rate also increases, but when the number of convolution kernels in the first layer increases to a certain number, the classification. The accuracy rate does not change or changes little, but the training time and model complexity increase. Therefore, under the condition of the number of samples in this paper, when actually designing the number of convolution kernels, the number of convolution kernels of the first convolutional layer is set to 8, and the number of convolution kernels of the second convolutional layer is set to 9. It can greatly reduce the complexity of the model while ensuring a higher accuracy.

## 20.3 Verification of Aeroengine Gas Path Fault Diagnosis Method Based on CNN and SVM

In order to verify that the model proposed in this paper has a good ability to identify aeroengine gas path failures, we made three sets of comparative experiments. The first set of experiments directly used support vector machines to classify the aeroengine gas path failures; the second set of experiments directly used convolutional neural networks identify aeroengine gas path faults; the third set of experiments uses SDAE and SVM-based gas path fault diagnosis models to identify aeroengine gas path faults. In this experiment, the SDAE parameters are set as follows: batch size is 10, the number of iterations is epoch = 500, the SDAE model structure is composed of an input layer, an intermediate layer, and an output layer. The number is 60, and the number of output layer nodes is 40.

The support vector machine kernel function uses a polynomial kernel. The experimental data set comes from the aeroengine gas path state data set collected in Chapter 3. Each experiment is repeated 10 times, and the average value of the experimental results is taken as the final fault recognition rate. The experimental results are shown in Table 3.1.

It can be seen from Table 20.2 that the training accuracy of the four methods on the training set can reach 100%, but there is a significant difference in the test

**Table 20.2** Comparison of failure recognition rates of different models

| Model | Training | prec (EGTindex) (%) | prec (TATindex) (%) | prec (HPT_Blade) (%) | prec (Normal) (%) | Diagnosis correct number/Total number of samples (%) |
|---|---|---|---|---|---|---|
| SVM | 1 | 50 | 80 | 50 | 89.7 | 80.6 |
| CNN | 1 | 0 | 0 | 0 | 100 | 69.39 |
| SDAE + SVMM | 1 | 85 | 95 | 91.3 | 93 | 92.2 |
| CNN + SVM | | 98 | 92 | 96 | 93.22 | 93.6 |

accuracy on the test set. The CNN method has the worst test accuracy and cannot detect the three types of faults. This is because the number of fault samples is too small, which leads to overfitting problems when training the CNN, and once again illustrates the importance of the number of samples for the end-to-end CNN method. Secondly, the test accuracy of the SVM method is also very poor, especially for EGT indication failure and HPTB failure. This is because of the redundancy between the parameter attributes in the original sequence samples, which results in a poor SVM classification effect. Compared with the first two methods, the SDAE-SVM method and CNN-SVM method have better fault diagnosis capabilities and can effectively detect three types of faults. It can be seen from Table 2.4 that the CNN-SVM method is superior to the SDAE-SVM method in identifying EGT indication faults and HPT_ blade faults, and the SDAE-SVM method is superior to CNN-SVM in identifying faults TAT SVM method. Therefore, in order to achieve more accurate fault diagnosis of aeroengine gas circuit, two methods can be used for fault diagnosis at the same time, and the results of the two methods can be mutually verified [5].

## 20.4 Conclusion

This paper designs a fault diagnosis algorithm based on the combination of CNN-SVM. Convert the aeroengine gas path state data into a time-parameter matrix; then, based on a large number of normal samples, use CNN to establish an engine gas path state feature extraction model; then use this state feature extraction model to perform feature mining on fault samples, through the features mined after multiple convolution and pooling operations are no longer directly connected to the classifier through the fully connected layer for classification but are used as the state features of the fault samples, and these features are classified using SVM. In order to make the proposed model have better classification effect, this paper also optimizes the structure parameters of CNN through a large number of experiments. Through three sets of controlled experiments, the effectiveness of the CNN-SVM method is fully proved. The experimental results show that the CNN-SVM-based aeroengine fault diagnosis method has a high overall diagnostic accuracy, and the detection rate of each type of fault is 90% the above. The CNN-SVM diagnosis method has broad application prospects in the engineering practice of aeroengine gas path fault diagnosis.

## References

1. Lecun, Y., Boser, B., Denker, J.S.: et al, pp. 396–404. Backpropagation Applied to Handwritten Cambridge, MA (1990)
2. Zip Code Recognition. Neural Comput **1**(4), 541–551 (1989)
3. Soren Bouma, One Shot Learning and Siamese Networks in Keras
4. https://sorenbouma.github.io/blog/oneshot/. Last accessed 2019/04/03

5. Cao, M., Li, L., Xie, H.: Application of improved genetic algorithm in microphone array optimization. J. National Univ. Defense Technol. **06** (2019)

# Chapter 21
# Improved Adaptive Particle Swarm Optimization Algorithm with a Two-Way Learning Method

**Yawen Wang, Qian Qian, Yong Feng, and Yunfa Fu**

**Abstract** Traditional particle swarm optimization algorithm has some disadvantages, such as slow convergence speed and easy to fall into local extremes. In order to improve the performance, an improved adaptive particle swarm optimization algorithm with a two-way learning method is proposed. First, the algorithm adaptively adjusts the algorithm according to the iteration periods of the optimization process. Specifically, the inertia weight and the value of the learning factor are changed nonlinearly, so as to better balance the search behavior of the particles in the group; Second, the idea of beetle search is introduced into the particle swarm algorithm to form a new two-way learning mechanism, which overcomes the limitations of the traditional particle swarm algorithm and help to increase the diversity of the population. In this way, the search scope is expanded, and the search accuracy of the algorithm is enhanced. Finally, the simulation is carried out on several multi-dimensional functions, and compared with other two related algorithms. The experimental results show that under the same experimental conditions, the improved algorithm has obvious advantages in optimization ability and convergence speed.

## 21.1 Introduction

Particle Swarm Optimization (PSO) is a swarm intelligence optimization algorithm first proposed by Eberhart and Kennedy [1] in 1995. Due to its simple operation and not being dependent on the gradient information of the target problem, this algorithm has been widely used in many fields once it is proposed, such as neural network classification [2] and image segmentation [3]. Although the PSO has good optimization performance, it cannot balance the global search and local search behaviors

Y. Wang · Q. Qian (✉) · Y. Feng · Y. Fu
Faculty of Information Engineering and Automation, Kunming University of Science and Technology, Kunming 650500, Yunnan, China
e-mail: qianqian_yn@126.com

Y. Wang · Q. Qian · Y. Feng
Yunnan Key Laboratory of Computer Technology Applications, Kunming University of Science and Technology, Kunming 650500, Yunnan, China

 
L. C. Jain et al. (eds.), *Smart Communications, Intelligent Algorithms and Interactive Methods*, Smart Innovation, Systems and Technologies 257,
https://doi.org/10.1007/978-981-16-5164-9_21

well during the optimization process [4], and it has the defects of "premature" and slow convergence. Therefore, many researchers have proposed different improvement methods. The literature [5] proposed an improved particle swarm algorithm, which can adaptively adjust the inertia weight $W$ in the algorithm update rule, which relieves the algorithm's tendency to "premature" convergence. The literature [6] proposes an asynchronous change strategy to update the learning factors in the PSO, which effectively improves the search ability of the algorithm; the literature [7] combines the Beetle Antennae Search Algorithm (BAS) with the PSO, which enriches the group diversity of the algorithm during the iteration process, and improves the convergence speed of the algorithm. The literature [3] introduced the simulated annealing mechanism into the standard PSO to enhance the local search capability of the algorithm.

The above improvements have improved the performance of the algorithm to a certain extent. This paper proposes a new PSO on the basis of previous studies. Specifically, to dynamically adjust both the inertia weight $W$ and learning factor $C$ of the PSO according to the iteration status of the algorithm. This dual adaptive method can better control the global and the local search behaviors of the algorithm; In addition, a two-way learning strategy that integrates the idea of beetle antenna search into the optimization process of the PSO is formed to improve adaptive particle swarm optimization algorithm (ABLPSO). The simulation test of this method shows that this method can better jump out of the local extremes, reduce the possibility of premature, and improve the performance of the algorithm.

## 21.2 Double Adaptive Optimization Strategy

In the standard PSO, the update of particles is mainly affected by the original speed, the individual historical optimal position, and the global optimal position. Among them, the size of the inertia factor $W$ determines the degree of influence from the particle's own speed, and the learning factors $C_1$ and $C_2$, respectively, determine the degree of the influence from the individual historical optimal and the global historical optimal. Therefore, the inertia weight $W$ and the learning factor $C$ are very important parameters in PSO. Literature [8] uses the Sigmoid function to adjust the inertia weight, and literature [9] uses a nonlinear method to adjust the learning factor $C$. On the basis of these previous studies, this paper proposes a new dual adaptive optimization strategy, in which both the inertia weight $W$ and the learning factor $C$ are adaptively adjusted.

### *21.2.1 Adaptive Inertial Weight*

The inertial weight $W$ is the key to maintaining the exploration and mining capabilities of the algorithm in PSO [10]. Appropriate adjustment of the $W$ can reduce the possibility of the algorithm falling into the local optimum. When the value of the $W$ is large, the particles have better global search ability, and when the value of $W$ is small, the local search ability is better. If the $W$ adopts a fixed value and the search ability of the particles remains unchanged, there will be a high probability of premature conditions. In addition, the optimization process of the PSO algorithm is complex and nonlinear [5], so this paper introduces a new adaptive inertia weight method, which nonlinearly adjusts $W$ according to the number of iterations of the particles. The aim is to achieve a certain balance between global search and local search. The update method of the inertia weight $W$ of the algorithm in this paper is shown in Formula (21.1):

$$W = W_{\min} + (W_{\max} - W_{\min})\frac{1}{1 + \exp(A(2\frac{t}{T} - 1))} \quad (21.1)$$

Among them: $W_{\min}$ is the minimum value of inertia weight, with a value of 0.1; $W_{\max}$ is the maximum value of inertia weight, with a value of 0.8; $A$ is a parameter that controls the curvature of the curve, after testing, $A$ is set to 6; $t$ is the current iteration number; $T$ is the maximum number of iterations.

The adaptive change of the inertial weight $W$ is shown in Fig. 21.1. At the beginning of the iteration, $W$ maintains a relatively stable and large state, so that most particles have better global search capabilities, thus improving the richness of the early population and avoiding premature phenomenon; in the later stage of the iterations, in order to improve the convergence ability of the algorithm, $W$ decreases rapidly, so that the particles can be updated with a small step size to achieve a more refined search.

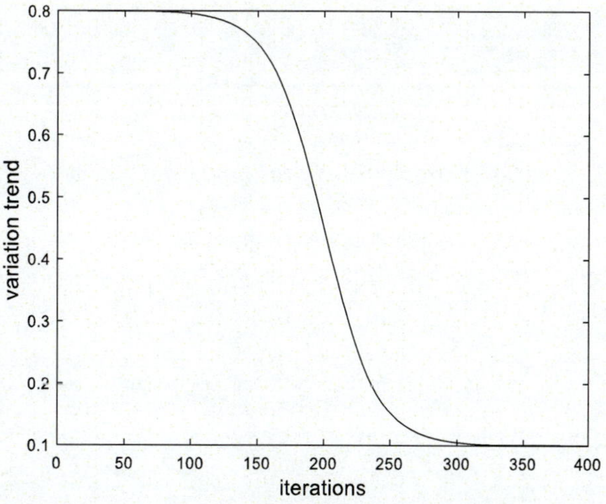


**Fig. 21.1** Inertial weight adaptive curve

### 21.2.2 Adaptive Learning Factor

The learning factor in the PSO represents the learning ability of the particles. The single-grain method is better in exploration and reduces the computational complexity [11], so the present algorithm uses a single-particle update method, which is to randomly select a particle $\mathrm{pbest}_k$ from the better particles as the learning object of each individual in the population. The update formula of learning factor $C$ is as follows:

$$C = C_{\min} + (C_{\max} - C_{\min})\left(1 - \left(\frac{\mathrm{fit}}{f_{\max}}\right)^3\right) \tag{21.2}$$

Among them: $C_{\min}$ is the minimum learning factor, with a value of 1.49445; $C_{\max}$ is the maximum learning factor, with a value of 2; fit is the fitness value of the current individual particle, and $f_{\max}$ is the largest fitness value among contemporary particles.

The learning factor in Fig. 21.2 shows a nonlinear downward trend. At the beginning of the algorithm iteration, the learning factor slowly decreases with a larger value, which makes the particles search at a larger speed in the solution space and increases the probability of the particles jumping out of the local optimum. In the later stage of the search, the rapid decrease of the learning factor can increase the convergence speed.

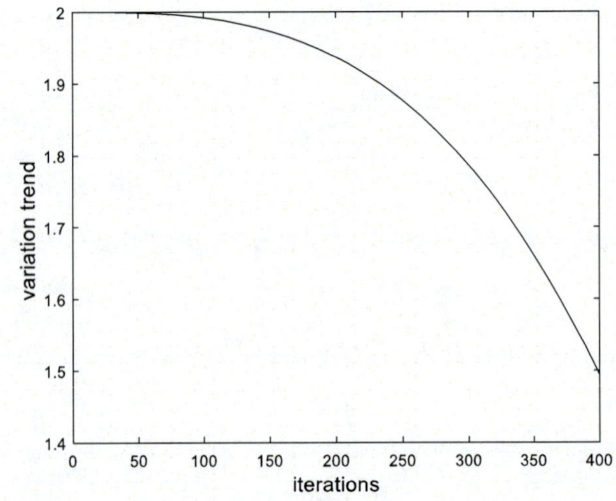


**Fig. 21.2** Learning factor $C$ adaptive curve

## 21.3 Basic Beetle Antenna Search Algorithm

The Beetle Antenna Search Algorithm (BAS) simulates the foraging process of longhorn beetle. When the beetle is looking for food, it judges the direction of the next move by sensing the intensity of the food odor. The odor intensity sensed by the left and right antennae of the beetle is an analogy to the fitness value in the function, and the specific orientation of the food is an analogy to the best point of the optimization function. The update of the position of the longhorn beetle in the algorithm is completed by calculating the fitness values $f(x_r)$ and $f(x_l)$ of the left and right antennae of the beetle, and then comparing the fitness of the left and right antennae. The left and right antenna positions of beetle are expressed as follows:

$$x_r = x + \vec{b}\,d \tag{21.3}$$

$$x_l = x - \vec{b}\,d \tag{21.4}$$

In the formula: $x$ is the current position of the beetle, $\vec{b}$ is a random unit vector, that is, the direction of the beetle is random during moving; $d$ is the distance from the beetle's antennae to the center of mass.

## 21.4 Two-Way Learning Strategy

This paper draws on the key ideas of the aforementioned BAS and proposes a two-way learning strategy. In the process of searching the solution space of the particles, each iteration will perform two behaviors: forward learning and reverse learning. That is, after selecting $\text{pbest}_k$ for each $x_i$, the solution learned from $\text{pbest}_k$ is calculated. In addition, a reverse solution is also calculated at the same time, and the better solution between the two solutions is chosen to improve the particle velocity and update the particle position. The speed and position update formula is:

$$v_i(t+1) = Wv_i(t) + \begin{cases} Cr_1(\text{pbest}_k(t) - x_i(t)) \\ Cr_1(x_i(t) - \text{pbest}_k(t)) \end{cases} \tag{21.5}$$

$$x_i(t+1) = x_i(t) + v_i(t+1) \tag{21.6}$$

Among them, $W$ is the inertia weight, and $C$ is the learning factor, $r_1$ is a random number between 0 and 1, $x_i(t)$ and $v_i(t)$ are the position and velocity of the particle, respectively, $\text{pbest}_k(t)$ is the historical optimal position of a randomly selected better individual $k$. As shown in the above formula, each time a particle is updated, it will perform two behaviors of forward learning and reverse learning at the same time according to Formula (21.5). Through learning, the particle acquires a new fitness and toward the direction of which producing a larger fitness value. This two-way learning particle update method not only expands the searching range of the algorithm, but also reduces the possibility of the algorithm falling into the local optimum to a certain extent.

## 21.5 Algorithm Steps

The process of the improved adaptive particle swarm algorithm based on two-way learning is as follows:

Step 1: Initialize the particle population, including the number of particles, algorithm parameters, and initial positions.
Step 2: Calculate the fitness function value corresponding to each particle position in the population, for each particle, sort the fitness values from excellent to poor, and randomly select a better individual $\mathrm{pbest}_k$ as the learning object of that particle.
Step 3: Update the inertia weight and learning factor adaptively according to Formula (21.1) and (21.2).
Step 4: First calculate the particle velocity by Formula (21.5), then calculate the corresponding solution according to Formula (21.6) and compare its fitness value, select the better one as the new particles.
Step 5: Update all particle fitness function values and global optimal positions.
Step 6: Judge whether the termination condition is satisfied, if satisfied, output the optimal solution, otherwise return to step 2.

## 21.6 Simulation Experiment and Result Analysis

### *21.6.1 Test Function*

In order to verify the performance of the algorithm, PSO, Adaptive Dynamic Change Particle Swarm Optimization Algorithm (ACPSO) [5], and ABLPSO proposed in this paper are used in the optimization of four representative complex functions. Table 21.1 shows the four benchmark function expressions, search space ranges,s and minimum values. $f_1$ and $f_2$ are single-peak functions, $f_3$ and $f_4$ are multi-peak functions.

### *21.6.2 Test Experiment and Analysis*

The experimental parameters of the three algorithms are: the population size of all algorithms is 100, and the maximum number of iterations is 400, $D$ is the dimension.

**Table 21.1** Test function

| Function | Formula | Range | Minimum |
|---|---|---|---|
| $f_1$ | $f_1(x) = \sum_{i=1}^{n} x_i^2$ | [−100, 100] | 0 |
| $f_2$ | $f_2(x) = \sum_{i=1}^{n} [x_i^2 + 0.5]^2$ | [−10, 10] | 0 |
| $f_3$ | $f_3(x) = \sum_{i=1}^{n} (x_i^2 - 10\cos(2\pi x_i) + 10)$ | [−5.12, 5.12] | 0 |
| $f_4$ | $f_4(x) = 20 + \mathrm{e} - 20\exp\left(-0.2\sqrt{\frac{1}{D}\sum_{i=1}^{D} x_i^2}\right) - \exp\left(\frac{1}{D}\sum_{i=1}^{D}\cos(2\pi x_i)\right)$ | [−32, 32] | 0 |

**Table 21.2** Number of adoption of reverse learning

| $M$ | Function | | | |
|---|---|---|---|---|
| | $f_1$ | $f_2$ | $f_3$ | $f_4$ |
| $D = 10$ | 32 | 33 | 29 | 30 |
| $D = 30$ | 21 | 19 | 16 | 18 |

In the PSO, $C_1 = C_2 = 1.5$, $W = 1.2$; in the ACPSO, $C_1 = C_2 = 1.5$, other parameters of the algorithm are set as that in reference [5]; the inertia weight $W$ and learning factor $C$ of the ABLPSO are set as that described at above. Let $M$ be the number of times that the reverse learning of the ABLPSO is accepted in the function optimization process, and this number for the above four functions at one execution is shown in Table 21.2. The data shows that the reverse learning method is indeed effective and being accepted many times when better solutions are found.

Table 21.3 shows the test results after each of the three algorithms is executed 30 times. It can be seen that the ABLPSO has obvious advantages in convergence speed compared to the PSO and ACPSO, and the optimal value of the solution is also closer to the ideal value of the function. Especially in the function $f_1$, the convergence speed of the ABLPSO is much faster than the other two algorithms. The functions $f_3$ and $f_4$ are multi-peak functions used to test the ability of the algorithm to jump out of the local extreme values. Although the PSO has a certain optimization ability and convergence speed when optimizing the multi-peak functions, the algorithm is easy to fall into the local optimum in the later stage. The ACPSO algorithm has better optimization ability than PSO. Compared with the other two algorithms, ABLPSO has the best convergence speed and ability to find the optimal solution.

**Table 21.3** Comparison of performance parameters

| Function | Algorithm | Mean convergence iteration | Average optimal value | The optimal value |
|---|---|---|---|---|
| $f_1$ | PSO | 60 | 1.2467e−01 | 4.3788e−05 |
| $D = 30$ | ACPSO | 45 | 2.1074e−05 | 3.3211e−08 |
| | ABLPSO | 29 | 7.5835e−10 | 1.4210e−10 |
| $f_2$ | PSO | 45 | 5.3634e+01 | 3.8208e−01 |
| $D = 30$ | ACPSO | 32 | 2.3560e−04 | 2.3560e−09 |
| | ABLPSO | 30 | 1.7792e−11 | 1.7792e−13 |
| $f_3$ | PSO | 156 | 2.7862e+01 | 3.2841e+01 |
| $D = 10$ | ACPSO | 100 | 5.8096e−01 | 2.9151e−09 |
| | ABLPSO | 51 | 3.3872e−05 | 2.6645e−15 |
| $f_4$ | PSO | 173 | 2.4795e+00 | 2.3103e−01 |
| $D = 10$ | ACPSO | 98 | 3.1194e−02 | 4.5745e−06 |
| | ABLPSO | 62 | 7.3328e−04 | 1.3172e−10 |

## 21.7 Conclusion

In order to solve the problem that the traditional particle swarm algorithm, such as slow convergence speed and easy to fall into local optimum, this paper proposes an ABLPSO algorithm. The algorithm introduces a two-way learning strategy to expand the population searching range and improve the global optimization capability, in addition, a dual adaptive optimization strategy is adopted to the inertia weight and learning factor of the algorithm to reduce the possibility of premature. Finally, through simulation experiments on four test functions, it is shown that the ABLPSO proposed in this paper has better convergence speed and optimization ability in function optimization than two other algorithms.

## References


1. Kennedy, J., Eberhartrc.: Particle swarm optimization. IEEE International Conference on Neural Networks, Perth, Australia: 1942–19489 (1995)
2. Hewahi, N.: A hybrid approach based on genetic algorithm and particle swarm optimization to improve neural network classification. J. Inf. Technol. Res. **10**(3), 48–68 (2017)
3. Wu, L., Cheng, W., Wang, X.: Application of simulated annealing particle swarm algorithm to optimize two-dimensional entropy image segmentation. Comput. Eng. Design **040**(009), 2544–2551 (2019)
4. Zhang, X., Kang, Q., Xia, W., et al.: Particle swarm optimization algorithm based on cross-reverse learning and cogranular social learning. J. Comput. Appl. **37**(011), 3194–3200 (2017)
5. Li, L., Zhang, X.: A new adaptive inertial weight chaotic PSO algorithm. Comput. Eng. Appl. **054**(009), 139–144 (2018)
6. Tong, Q., Zhao, Q., Li, M.: Particle swarm optimization algorithm based on adaptive dynamic change. Microelectron. Comput. **36**(2), 6–10

7. Wu, Q., Zhang, W., Yang, H. et al.: Application of particle swarm optimization based on longhorn whisker algorithm in PID parameter tuning. Electron. Sci. Technol. **33**, 369(06), 22–27 (2020)
8. Tian, D., Zhao, T.: Adaptive particle swarm optimization algorithm based on Sigmoid inertia weights. J. Comput. Appl. **28**(12), 3058–3061 (2008)
9. Xu, S., Xia, W., Dai, A.: A particle swarm optimization algorithm with improved learning factor. Inf. Secur. Technol. **003**(007), 17–19 (2012)
10. Ding, X., Wu, X., Cheng, H.: Research on wireless sensor network coverage problem based on improved particle swarm algorithm and feature point set. Chin. J. Electron. **44**(4), 967–973 (2016)
11. Zhang, X., Wang, X., Kang, Q., Cheng, J.: Differential mutation and novel social learning particle swarm optimization algorithm. Inf. Sci. (2018)

# Chapter 22
# Flower Pollination Algorithm Based on Beetle Antennae Search Method

**Jiayi Zhou, Qian Qian, Yunfa Fu, and Yong Feng**

**Abstract** In this paper, a new flower pollination algorithm based on beetle antennae search method (TBFPA) is proposed to deal with the slow convergence speed problem of traditional flower pollination algorithm. Specifically, after the new individuals are generated by the way of flower pollination algorithm, the beetle antennae search method is used to search the nearby solution space of the new individuals, thus increasing the probability to reach better solutions and speeding up the convergence speed of the algorithm. In addition, a new uniform distribution sampling method is used to generate the initial population so as to spread the initial solutions all over the solution space. In order to verify the effectiveness of TBFPA, different standard functions are tested between TBFPA and FPA algorithms and the results proved that TBFPA has faster convergence speed and stronger global searchability.

## 22.1 Introduction

Flower pollination algorithm (FPA) is a new swarm intelligence optimization algorithm proposed by Professor Yang of Cambridge University in 2012 [1]. FPA has the characteristics of less parameters, easy adjustment and easy implementation. Thc algorithm relies on probability parameter $P$ to dynamically control the conversion between global search and local search of the solutions. Levy flight is used in global search of the algorithm, which makes it have good global optimization ability. Many researchers have studied, improved and applied the FPA algorithm: Zhou et al. [2] and others used elite reverse learning to initialize the population, which improved the diversity of the initial population; in reference [3], Xiao Huihui used the gravity between two individuals of the population and Levy flight to update the individual positions. In reference [4], the simplex method was used in the local optimization

J. Zhou · Q. Qian (✉) · Y. Fu · Y. Feng
Faculty of Information Engineering and Automation, Kunming University of Science and Technology, YunnanKunming 650500, China
e-mail: qianqian_yn@126.com

Yunnan Key Laboratory of Computer Technology Applications, Kunming University of Science and Technology, YunnanKunming 650500, China

 
L. C. Jain et al. (eds.), *Smart Communications, Intelligent Algorithms and Interactive Methods*, Smart Innovation, Systems and Technologies 257,
https://doi.org/10.1007/978-981-16-5164-9_22

part to expand or compress the poor individuals before them entering the next iteration, so as to improve the local optimization ability of the FPA. The above literatures have improved the optimization ability of FPA to a certain extent, but many improved algorithms still have the problems of high complexity and slow running speed.

Beetle antennae search algorithm (BAS) [5] is a novel bio-inspired intelligent optimization algorithm proposed by Jiang in 2017. The idea comes from the simulation of longicorn foraging behavior. Compared with particle swarm optimization, flower pollination algorithm and other swarm intelligence optimization algorithms, BAS only needs one individual, so it can be integrated with various swarm intelligence optimization algorithms without great increasing of the algorithm complexity. Chen et al. [6] proposed PSO based on BAS and applied it to the portfolio model with complete cost, which improved the effectiveness and stability of PSO. Zhao and Qian [7] proposed a chaotic beetle swarm algorithm to improve the convergence and optimization ability of BAS. In this paper, the single individual of BAS is extended to a population, and the optimization ability of the original algorithm is improved.

In this paper, beetle antennae search method from BAS is combined with flower pollination algorithm. Specifically, BAS method is used to evaluate the nearby solution space of each new individual that was calculated by FPA, thus increasing the convergence speed and the probability to get the global optimal solution. In addition, a new initialization method is proposed to accelerate the convergence speed in the early stage of the algorithm and improve the ability of the algorithm to jump out of the local optimal solution in the later stage. The effectiveness of the improved algorithm (called "TBFPA") is verified by experiments.

## 22.2 Basic Principles

### *22.2.1 Flower Pollination Algorithm (FPA)*

FPA randomly initializes a population $X_i^t$ with NP individuals, where $X_i^t = \left[x_{i,1}^t, x_{i,2}^t, \ldots, x_{i,j}^t, \ldots, x_{i,D}^t\right]$, $i = 1, 2, \ldots, \text{NP}, j = 1, 2, \ldots, D$, $D$ denotes the dimension of the optimization problem, and $T$ represents the current iteration number of the population [1]. After initializing the population, FPA evaluates the fitness value of each individual and selects the individual with the highest fitness value as the global optimal solution. Then FPA depends on the probability $P$ to decide searching globally or locally until the termination condition is reached. The global search operation is defined as follows:

$$X_i^{t+1} = X_i^t + L(\lambda)\big(\text{gbest} - X_i^t\big) \tag{22.1}$$

Among them, gbest is the current global optimal solution, $L(\lambda)$ is a random real number obeying Levy distribution with index $\lambda$, and the value of $\lambda$ recommended in reference [1] is 1.5.

The local search operation is defined as follows:

$$X_i^t = X_i^t + \varepsilon \cdot \left(X_j^t - X_k^t\right) \tag{22.2}$$

where $\varepsilon$ is a random real number with uniform distribution between [0, 1], and $j$, $k$ are two integers that are different from each other in [1, NP].

The FPA algorithm first generates the initial population randomly. At each iteration, each individual performs the global search with probability $p$ as shown in Formula (22.1) or the local search with probability $1 - p$ as shown in Formula (22.2). After the search is completed, FPA evaluates the new position of the individual and updates it if the new position is better than the original position.

### 22.2.2 *Beetle Antennae Search Algorithm (BAS)*

The basic principle of BAS [4] is to simulate the process of beetle to determine its own flight direction through the perception of food taste by left and right antennae without knowing the location of foods. If the taste intensity of the left antennae is stronger than that of the right antennae, it will fly to the left; otherwise, the beetle will fly in the direction of the right antennae. The modeling process of BAS is as follows:

The positions of the left and right antennae of beetle were as follows:

$$\begin{cases} x\ \text{left} = x + L \cdot \overrightarrow{b} \\ x\ \text{right} = x - L \cdot \overrightarrow{b} \end{cases} \tag{22.3}$$

where $x$ is the current solution position, $L$ is the initial antennae length, which decreases with the number of iterations, and $\overrightarrow{b}$ is a randomly generated $D$-dimensional unit vector, which represents the direction of the beetle. In the process of calculation, the direction of left antennae of beetle was defined as positive direction. The fitness function values of left and right antennae of beetle were evaluated, and the next position of beetle was updated by the following formula:

$$X_i^{t+1} = X_i^t - \text{step} \cdot \overrightarrow{b} \cdot \text{sign}(f(x\ \text{left}) - f(x\ \text{right})) \tag{22.4}$$

where step is the search step size, which decreases with the number of iterations. $f(x)$ is the fitness function. sign() is a symbolic function. In the problem of finding the minimum value, if $f(x\ \text{left}) > f(x\ \text{right})$, which means $x$ right is better than $x$ left, then the value of sign() is 1, and beetle moves to the right.

### 22.2.3 Population Initialization Method of Uniform Sampling Using Toeplitz Matrix

Literature [1] shows that the convergence of swarm intelligence algorithm is greatly affected by the initial population distribution. In this paper, a method is proposed to divide each dimension of solutions into several sections, and then initial solutions are generated from going through each section to ensure that each section of each dimension is covered by an individual.

Toeplitz matrix, abbreviated as T-type matrix, is characterized in that each element is equal to the element in the upper left corner except for the elements in the first row and the first column. Therefore, only the first row and the first column are needed to generate the entire matrix. The Toeplitz matrix is shown in Fig. 22.1. The special Toeplitz matrix can make each column of the matrix be a different arrangement of the same elements. The population initialization steps are as follows.

Step 1: Each dimension of $D$-Dimension is divided into NP segments, and NP is the population size.
Step 2: Generating Toeplitz matrix $T$. The first line is a sequence of consecutive integers of $[1, D]$. The first element in the first column is 1, and the second to NP elements is the inverse sequence of integers [NP, 2]. Each column of the resulting matrix is a different arrangement of the same elements. The number of rows of the matrix is equal to the population size, and the number of columns of the matrix is equal to the dimension of the problem. Take the sequence generated by $D = 5$ and $\mathrm{NP} = 10$ as an example. As shown in Fig. 22.2, each column is a non-repeating integer sequence of [1, NP].
Step 3: For the $i$-th individual, the value of the $j$-th dimension is randomly selected from the $T$ $(i, j)$-th section of the $j$-th dimension. Suppose that the $T$-matrix generated by step 2 is shown in Fig. 22.2 ($D$ = 5, NP = 10). For the fourth individual, its third dimension value is randomly extracted from the $T$ (4, 3) = 10-th section of the third dimension. If the value range of the third dimension is [0, 10], then the tenth segment is [9, 10], and the third dimension of the fourth individual is a random number between [9, 10]. Because every column of $T$-matrix is a non-repeating integer sequence of [1, NP], this initialization method ensures that there is an individual in each uniform section of each dimension, which makes the initial population more uniform and diverse.

**Fig. 22.1** Toeplitz matrix

$$T = \begin{pmatrix} t_0 & t_1 & t_2 & \cdots & t_{n-1} \\ t_{-1} & t_0 & t_1 & \cdots & t_{n-2} \\ t_{-2} & t_{-1} & t_0 & \cdots & t_{n-3} \\ \vdots & \vdots & \vdots & \ddots & \vdots \\ t_{-n+1} & t_{-n+2} & t_{-n+3} & \cdots & t_0 \end{pmatrix}$$

**Fig. 22.2** *T*-Matrix with $D = 5$, NP = 10

| | | | | |
|---|---|---|---|---|
| 1 | 2 | 3 | 4 | 5 |
| 10 | 1 | 2 | 3 | 4 |
| 9 | 10 | 1 | 2 | 3 |
| 8 | 9 | 10 | 1 | 2 |
| 7 | 8 | 9 | 10 | 1 |
| 6 | 7 | 8 | 9 | 10 |
| 5 | 6 | 7 | 8 | 9 |
| 4 | 5 | 6 | 7 | 8 |
| 3 | 4 | 5 | 6 | 7 |
| 2 | 3 | 4 | 5 | 6 |

### *22.2.4 Flower Pollination Algorithm Based on Beetle Antennae Search Algorithm (TBFPA)*

Flower pollination algorithm balances the process of global search and local search by probability *P* and jumps out of local optimum value by Levy flight. By evaluating the solution space of the left and right antennae directions of the individual, the BAS updates the solution to the better direction of the two, and the convergence speed. The idea of TBFPA is to guide individuals to move toward the global optimal solution by FPA, and then the solution space near the individual's new position evaluated and updated by the method of BAS. Based on the results of many experiments, when the length of antennae d and search step of TBFPA are small and fixed values, the convergence speed is most accelerated. The small value of parameters is useful to evaluate the nearby solution space of the individual's position, while the fixed value of parameters is used to ensure that the algorithm has the ability to evaluate the solution space during the whole processing period. The global search and update formula of TBFPA is as follows:

$$\begin{cases} x\ \text{left} = x_i^t + L(\lambda)\left(\text{gbest} - x_i^t\right) + d \cdot \overrightarrow{b} \\ x\ \text{right} = x_i^t + L(\lambda)\left(\text{gbest} - x_i^t\right) - d \cdot \overrightarrow{b} \end{cases} \tag{22.5}$$

$$x_i^{t+1} = x_i^t + L(\lambda)(\text{gbest} - x_i^t) - \text{step} \cdot \overrightarrow{b} \cdot \text{sign}(f(x\ \text{left}) - f(x\ \text{right})) \tag{22.6}$$

The local search and update formula of TBFPA are as follows:

$$\begin{cases} x\ \text{left} = x_i^t + \varepsilon\left(x_j^t - x_k^t\right) + d \cdot \overrightarrow{b} \\ x\ \text{right} = x_i^t + \varepsilon\left(x_j^t - x_k^t\right) - d \cdot \overrightarrow{b} \end{cases} \tag{22.7}$$

$$x_i^{t+1} = x_i^t + \varepsilon\left(x_j^t - x_k^t\right) - \text{step} \cdot \overrightarrow{b} \cdot \text{sign}(f(x\ \text{left}) - f(x\ \text{right})) \tag{22.8}$$

The process of TBFPA is as follows:

Step 1: Initialing dimension $D$, population size NP, antennae length $d$, search step and global search probability $P$.
Step 2: Initialing population by the method in 2.3.
Step 3: Calculating the fitness values of each individual and saving the global optimal solution, $i = 1$, $t = 1$.
Step 4: If the algorithm satisfies the termination condition, the global optimal solution is outputted; otherwise, goes to step 5.
Step 5: Generating rand (0, 1) and comparing it with $P$. If it is larger than $P$, goes to step 6, otherwise goes to step 7.
Step 6: According to Formula (22.6), a global search is executed to generate the next generation of individuals $X_i^{t+1}$ and goes to step 8.
Step 7: According to Formula (22.8), the next generation of individuals $X_i^{t+1}$ is generated by local search.
Step 8: If $X_i^{t+1}$ is better than $X_i^t$, then $X_i^{t+1}$ enters the next generation, otherwise $X_i^t$ enters the next generation, $i = i + 1$. Goes to Step 4.

## 22.3 Experiment

### 22.3.1 Test Functions and Parameter Settings

Two standard test functions are introduced to compare TBFPA, FPA and BASFPA. BASFPA is the same as TBFPA with the difference that it uses traditional random initialization method. The population size of all populations was 50, the flower pollination probability $P$ was 0.8, the antennae length of beetle was 0.1, the search step size of beetle was 0.1, the iteration times were 2000, the problem dimension was set to 20, and all functions were run 20 times (Table 22.1).

**Table 22.1** Test function

| Function name | Function | Range | Optimal value |
|---|---|---|---|
| Shpere | $f_1 = \sum_{i=1}^{D} x_i^2$ | (−50, 50) | 0 |
| Griewank | $f_2 = \frac{1}{4000}\sum_{i=1}^{D} x_i^2 - \prod_{i=1}^{D} \cos\left(\frac{x_i}{\sqrt{i}}\right) + 1$ | (−50, 50) | 0 |

**Table 22.2** Different algorithms calculate the average optimal fitness value of the function

| Function | Algorithm | Average optimal value |
|---|---|---|
| $F_1$ | FPA | 0.22623 |
| | BASFPA | 0.10039 |
| | TBFPA | 0.03202 |
| $F_2$ | FPA | 0.23422 |
| | BASFPA | 0.10040 |
| | TBFPA | 0.05171 |

### 22.3.2 Experimental Results and Analysis

The algorithms run the two functions independently for 20 times each, and the calculated average fitness values are shown in Table 22.2. The average optimal fitness value reflects the final convergence accuracy of the algorithm. It can be seen from the table that BASFPA improved the convergence accuracy and the searching ability of FPA algorithm. The TBFPA with a uniform initial population further increased convergence speed and the convergence ability, compared with BASFPA.

The convergence curves of the algorithm for the two functions are shown in Figs. 22.3 and 22.4. The ordinate is the fitness values of the optimal individual of the population, and the abscissa is the iteration number of the population. It can be seen from the figure that the convergence speed and accuracy of TBFPA are higher than

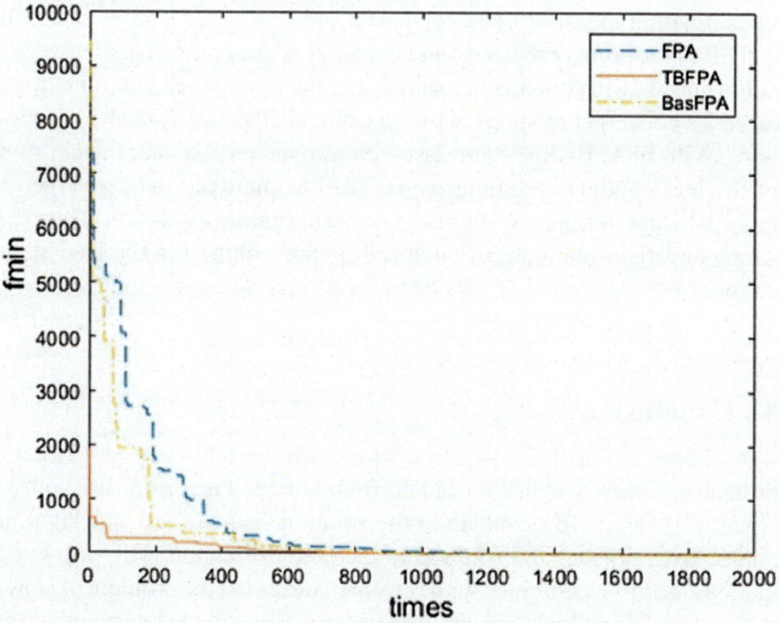


**Fig. 22.3** Convergence curves of three algorithms for Shpere

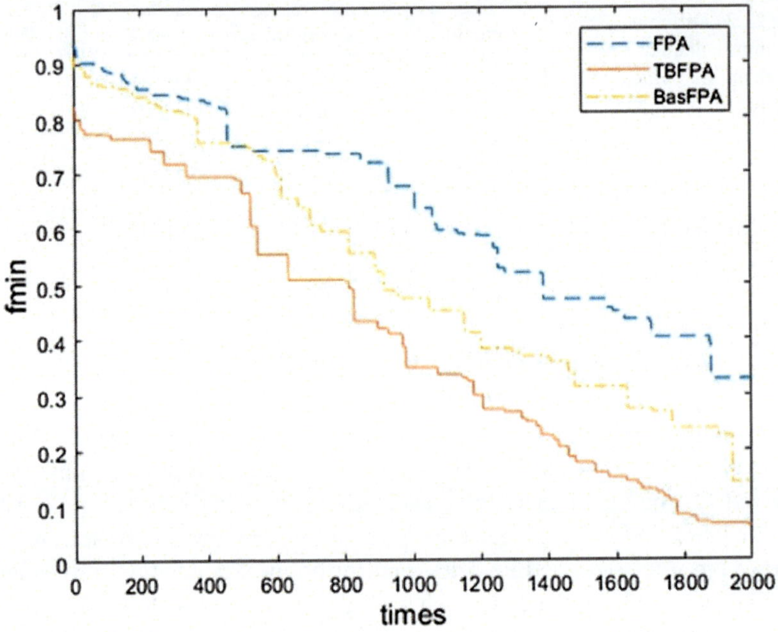


**Fig. 22.4** Convergence curves of three algorithms for Griewank

that of FPA for both functions $F_1$ and $F_2$, which indicates that the improved strategy can improve the searching ability of FPA algorithm. For function $F_1$, the improved flower pollination algorithm has better convergence speed and precision than the original FPA algorithm in the early and late stages. The comparison between TBFPA algorithm and BASFPA algorithm shows that the new initialization method greatly improved the convergence speed of the algorithm in the initial stage. For function $F_2$, compared with FPA, BASFPA has faster convergence speed and the ability to jump out of the local optimal solution; while TBFPA improved the convergence speed in the initial stage because of its more uniform initial population distribution and has strong ability to jump out of the local optimal solution in the later stage of the algorithm.

## 22.4 Conclusion

Beetle antennae search algorithm and flower pollination algorithm are both excellent intelligent algorithms. By evaluating the solution space of left and right antennae directions, BAS updates the solution to the better direction. The improved flower pollination algorithm combined with BAS method solves the problem of slow convergence speed of the original flower pollination algorithm in the later stage. In addition, the new initial population method ensures the diversity of the initial population, thus further improved the performance of the algorithm. The experimental results show that the convergence speed and the ability to jump out of the local optimal solution of TBFPA are better than FPA. In the future research, we will consider adding the disturbance operation to the population to maintain the diversity of the population, so that the algorithm can continue to have the ability to jump out of the local optimum.

## References


1. Yang, X.S.: Flower pollination algorithm for global optimization. Unconventional Computation and Natural Computation. Lecture Notes in Computer Science, vol. 7445, pp. 240–249 (2012)
2. Zhou, Y., Wang, Luo, Q.: Elite opposition-based flower pollination algorithm. Neurocomputin **188**, 94–310 (2016)
3. Xiao, H., Wan, C., Duan, Y., Tan, Q.: Flower pollination algorithm based on gravity search mechanism. J. Automatica Sinica **43**(4), 576–594 (2017)
4. Xiao, H.: A flower pollination algorithm based on simplex method and self-adaptive step. Comput. Eng. Sci. **38**(10), 2126–2133 (2016)
5. Jiang, X.Y., Li, S.: BAS: Beetle antennae search algorithm for optimization problems. Int. J. Robot. Control **1**(1), 1–4 (2018)
6. Chen, T., Yin, H., Jiang, H.: Particle Swarm optimization algorithm based on Beetle Antennae search for solving portfolio problem. Comput. Syst. Appl. **28**(2), 171–176 (2019)
7. Zhao, Y., Qian, Q.: Novel Chaos Beetle Swarm searching algorithm with learning and competitive strategies. Commun. Technol. **51**(11), 60–66 (2018)

# Chapter 23
# A Video Transmission Control Strategy in Wireless Network

**Run Zhou, Min Xiang, and Yu Liu**

**Abstract** Concerning the problem of poor video transmission quality caused by wireless network fluctuation, proposes a video transmission strategy based on fountain code. Combined with the characteristics of H.264 video stream, fountain code is used as the video transmission application layer coding scheme; a rate-based bandwidth estimation model is constructed, and the encoding parameters of fountain code are dynamically adjusted according to the bandwidth estimation results. Experiments show that the proposed video transmission control strategy can effectively improve the transmission quality of video data in wireless network environment with high packet loss rate. It has high coding flexibility and can effectively improve the bandwidth utilization.

## 23.1 Introduction

With the popularization of 4G network and Wi-Fi, the traffic proportion of video transmission services in the mobile Internet increases explosively [1]. How to improve the quality of video transmission services has also become the focus of research. However, the current wireless network cannot effectively adapt to the video transmission service in each application scenario. The influence of wireless network fluctuation on video transmission quality is an urgent problem to be solved.

Fountain code as a forward error-correcting code technology, through the continuous generation of coding data packets in the data source, only need to receive a certain number of coding data packets can recover the original data. These advantages make fountain code become an excellent solution to improve the quality of wireless network. Reference [2] introduced LT codes and proposed ideal soliton distribution and robust soliton distribution. Reference [3] introduced the application of fountain codes in wireless optical communication. Reference [4] optimized the degree distribution of LT codes, reduced the decoding overhead and complexity makes it more suitable for underwater acoustic communication. Reference [5] constructed a new LT

R. Zhou (✉) · M. Xiang · Y. Liu
Key Laboratory of Industrial Internet of Things and Networked Control, Chongqing University of Posts and Telecommunications, Chongqing, China

 
L. C. Jain et al. (eds.), *Smart Communications, Intelligent Algorithms and Interactive Methods*, Smart Innovation, Systems and Technologies 257,
https://doi.org/10.1007/978-981-16-5164-9_23

codes distribution through the comprehensive Poisson distribution, which improved the reliability and effectiveness of communication in the cognitive radio link.

Most of the above optimizations for LT codes are based on the degree distribution optimization for specific communication scenarios, without considering the optimization the optimization of coding strategies in practical use. In view of the above problems, this paper uses fountain code and bandwidth measurement algorithm to construct a video transmission control strategy, determines the current wireless network state through bandwidth measurement technology, and adjusts the LT codes coding redundancy.

## 23.2 Video Transmission Control Strategy

At present, the video service based on streaming media technology has become the mainstream. However, in practical applications, it is difficult to achieve good results in different application scenarios only using streaming media protocol for transmission, which often needs to be optimized in combination with application scenarios. In the network with high packet loss rate, it is important that the data can reach the receiver accurately. The video data are encoded and processed as the input symbols of LT codes, and then transmitted. Even if the original video data are lost in the transmission process, the decoder can also recover the original data from the set of generated coding symbols, avoiding data retransmission. Therefore, using LT codes to construct video transmission strategy can effectively reduce the impact of packet loss on video data, so that the quality of video stream in the network with large packet loss rate can be improved effectively.

### *23.2.1 Data Coding Strategy Based on LT Codes*

LT codes are the first practical fountain code, and the number of coded symbols can be generated as needed. They are sent through the channel until enough coded symbols arrive at the decoder, then the decoding is completed and the recovery of lost data is realized. Since the decoder can recover data from almost the smallest number of coded symbols, LT codes are close to the best relative to any erasing channel. Reference [2] proposed two-degree distributions of LT codes, namely ISD and RSD. The ISD is a distribution that satisfies by Eq. (23.1).

$$\rho(i) = \begin{cases} 1/k, & i = 1 \\ 1/(i(i-1)), & i = 2, \ldots, k \end{cases} \tag{23.1}$$

In Eq. (23.1), $k$ represents the number of input symbols, $k \in N$, $N$ represents a positive integer, $i$ represents the number of input symbols used for encoding selected from each input symbol, called degree, $i \in [1, k]$, $\rho(i)$ represents the ISD.

The ISD is easy to lose the output symbol of degree-1 in practical use, resulting in decoding interruption. In order to overcome this defect, based on the ISD, Reference [2] proposed RSD with stronger practical application ability, which can be expressed by Eqs. (23.2) and (23.3).

$$\mu(i) = [\rho(i)+\tau(i)]/Z, i = 1, \ldots, k \tag{23.2}$$

$$\tau(i) = \begin{cases} R/(i * k), & i = 1, \ldots, k/R - 1 \\ R \ln(R/\delta)/k, & i = k/R \\ 0, & i = (k/R) + 1, \ldots, k \end{cases} \tag{23.3}$$

In Eqs. (23.2) and (23.3), $R = c \cdot \ln(c/\delta) \cdot \sqrt{k}$ represents the number of output symbols of degree-1, $c > 0$ and $\delta \in [0,1]$, $Z = \sum_{i=1}^{k} [\rho(i) + \tau(i)]$. $\mu(\cdot)$ represents the RSD.

H.264 Video Streams $\{R - 1\}$ can be regarded as composed of multiple group of picture (GOP). Each GOP is composed of a certain number of I-frames, B-frames, and P-frames. I-frame is a complete image and does not need to refer to other frames when decoding. While B-frame and P-frame are predictive frames, which only store the motion vector rather than the complete image and need to refer to I-frame when decoding. In addition, the reference I-frame in the decoder is refreshed whenever a new I-frame arrives in the buffer, thus avoiding the spread of B-frame and P-frame decoding errors. Therefore, when encoding H.264 Video Streams, a group of GOP can be used as an independent input symbol unit, and LT codes encoding can be carried out combined with the RSD to obtain the encoded data package. In order to ensure the decoding success, there will be coding redundancy, which can be expressed by Eq. (23.4).

$$\varepsilon = \frac{K - k}{k} \tag{23.4}$$

In Eq. (23.4), $K$ represents the data packet encoded by the input symbol, $k$ represents the input symbol, and $\varepsilon$ represents the coding redundancy.

As the packet loss rate increases, the loss of correlated coded data packets will lead to an increase in the probability of decoding failure. If the video data is encoded by the fixed coding redundancy, where a fixed code redundancy is applied for video data encoding, it is necessary to adopt a large data redundancy for the purpose of successful decoding of video data. This will result in waste of resources and reduction of efficiency when the packet loss rate is low. Therefore, when using LT codes to encode the video data, the coding redundancy should be determined according to the bandwidth situation, so as to realize the adaptation of coding redundancy and network bandwidth and improve the coding flexibility and bandwidth utilization of LT codes.

### 23.2.2 Code Adaptive Coding Redundancy

At present, bandwidth measurement methods can be divided into active measurement and passive measurement according to whether it is necessary to inject data packets into the network. The active measurement method requires continuous injection of additional data packets into the network, which will increase the link burden in network congestion. The detected data packets cannot be guaranteed to be reliable with UDP, so it is difficult to accurately measure the available bandwidth of the network. However, the video stream transmission has the characteristics of large amount of data and long duration, which enable us to estimate the bandwidth value by measuring the video stream directly. Reference [6] proposed a model to estimate the available bandwidth by calculating the GOP sending time of each group of video files. This algorithm can simply and quickly estimate the current network bandwidth by sending rate. This paper will combine this algorithm to construct LT code redundancy adaptive strategy.

From the previous section, a GOP is a coding input unit, so the available bandwidth of wireless network can be estimated by the sending rate of a GOP. The size of a GOP is expressed as $S_j$, the sending time is $T_j$, $j$ is the number of the current GOP, the bandwidth value $B_j$ can be calculated by Eq. (23.5).

$$B_j = \frac{S_j}{T_j} \tag{23.5}$$

In Eq. (23.5), a bandwidth can be estimated according to each GOP sent. Although the redundancy of fountain codes is not directly related to the bandwidth value, it shows a certain positive correlation with the packet loss rate. Therefore, it is necessary to establish the relationship between the bandwidth value and the packet loss rate.

When the estimated bandwidth is greater than the video bit rate, it is considered that the current bandwidth meets the requirements of video stream transmission, and the packet loss probability is small. When the estimated bandwidth value is not greater than the video stream bit rate, it can be considered that the current bandwidth quality is poor and cannot satisfy the video stream transmission, and the probability of packet loss increases. Therefore, the relationship between coding redundancy $\varepsilon$ and available bandwidth of wireless networks can be expressed by Eq. (23.6).

$$\varepsilon = \begin{cases} a \cdot (V - B), & V - B > 0 \\ 0, & V - B < 0 \end{cases} \tag{23.6}$$

In Eq. (23.6), $V$ represents the video stream rate, $B$ represents the available bandwidth of the current wireless network, and $a$ represents the proportional coefficient. According to experimental, $a$ equals approximately 3. When the video bit rate is less than the current bandwidth value, it can be transmitted directly without additional coding redundancy. When the video rate is greater than the bandwidth value, it is determined that the current bandwidth is insufficient and the packet loss probability is large. The coding redundancy of LT codes is set according to the difference between the bandwidth value and the video rate. Therefore, the steps of video transmission control strategy are as follows.

## 23.3 Experimental Results

The test environment of this paper is as follows: an ARM embedded device as a video stream receiver, a PC as a video stream transmitter, through the LAN communication. In the PC side, the Linux kernel function module traffic control (tc) is used to control the packet loss rate and simulate the complex wireless network environment.

Select a video of 2046128 bytes for the experiment. Compare the receipt of data at receiving end under the circumstances of direct transmission and transmission after encoding using LT code in the network environments of different packet loss rates and reference [7] sets $c = 0.05$, $\delta = 0.05$. The numbers of received bytes are as shown in Table 23.1.

It can be seen from Table 23.1 that when the LT codes are not used to encode the video data, due to the lack of compensation for data packet loss, the number of packet loss increases gradually as the packet loss rate increases. The video data is processed by LT codes, even if there are different degrees of packet loss in the network, the receiver can still decode the encoded data packet and recover the video data, which significantly reduces the impact of network packet loss on the video. In addition, when the packet loss rate reaches 5%, the direct transmission of video in the receiving end will incur obvious frame loss, jitter, and screen more serious. In the network environment with 5% packet loss rate, the video processed by LT codes can still be smoothly played, and the video recovery effect is good.

Fixed coding redundancy 0.5 and 1 are set, respectively, to compare with the proposed adaptive coding redundancy, as shown in Fig. 23.1.

It can be seen from Fig. 23.1 that when the packet loss rate increases to 5%, the decoding success rate of the video data encoded by the coding redundancy 0.2 is less than 0.8 at the receiving end, while the decoding success rate of the coding group with coding redundancy 0.5 is significantly decreased when the packet loss rate is

**Table 23.1** Video data receiving bytes statistics table

| Packet loss rate | 0.5% | 1% | 5% |
|---|---|---|---|
| UDP | 2036328 | 2022328 | 1953728 |
| LT encode | 2045400 | 2044000 | 2034200 |

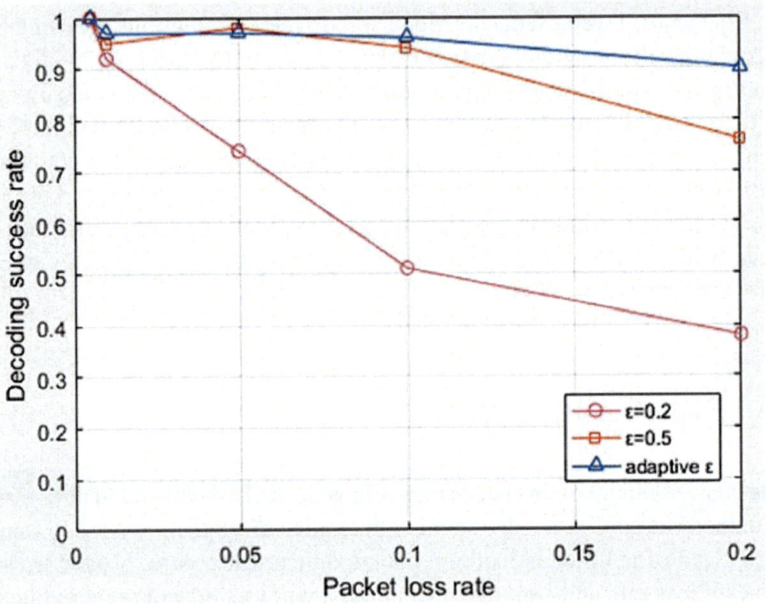


**Fig. 23.1** Comparison of decoding success rates of different $\varepsilon$ values

higher than 10%, which is not suitable for the environment with high packet loss rate. In the packet loss rate test environment, the proposed adaptive coding redundancy has high decoding success rate.

## 23.4 Conclusion

This paper proposes a video transmission control strategy suitable for wireless network environment with large packet loss rate. Specifically, it uses LT code to encode the video stream to reduce the impact of packet loss on the video stream and establishes a bandwidth estimation model on the basis of GOP transmission rate to achieve the coding redundancy adaptive of LT code, promoting the flexibility of LT Code and improving the transmission performance of video stream in the environment with large packet loss rate.

## References


1. Li Rong, X., Ri Jing, Y., Xin, J.: Sensor-based rate control method for mobile streaming media. Comput. Sci. **45**(10), 124–129 (2018)
2. Luby, M.: LT Codes. In Proceedings of the 43rd Annual IEEE Symposium Foundations of Computer Science (FOCS). IEEE Press, Vancouver (2002)

3. Yi, W., Hong-Zhan, L., Yuan, H., et al.: Development status and prospect of fountain codes in wireless optical communication. Optoelectron. Eng. **47**(03), 48–57 (2020)
4. Ya-Chen, L., Peng-Cheng, W., De-Hong, T., et al.: Degree distribution optimization of LT codes. J. Northwest Univ. Technol. **38**(03), 627–633 (2020)
5. Ben-Shun, Y., Wei-Qing, Y.: Improvement of LT code distribution and its application in cognitive radio link maintenance. J. Commun. **39**(04), 76–83 (2018)
6. Chang-Hao, W., Jian, W., Li-Hua, F., et al.: Adaptive video stream selection algorithm for cloud media networks based on prediction mechanism. Comput. Appl. Res. **33**(11), 3426–3429 (2016)
7. Ling, Y., Shi-Li, S., Guo-Chao, L., Xiao, C., et al.: Performance analysis and simulation of LT codes. Commun. Technol. **45**(05), 1–3 (2012)

# Chapter 24
# Data Correction-Based Data Fusion Method for Internet of Things Terminals

**Rujie Lei, Min Xiang, Peng Wang, Ruiheng Ma, and Kun Yu**

**Abstract** Accurate and reliable sensor data are essential for the Internet of Things (IoT) devices to operate well. To improve the accuracy of sensor data, a data correction-based data fusion method for IoT terminals was proposed. First, each terminal used the dynamic time warping (DTW) algorithm to calculate the distance between the sensor data. Then, the Spatial-temporal correlation degree was calculated by using the exponential function. Finally, the weighted data fusion method was used based on the Spatial-temporal correlation degree. The fusion result of each terminal was used as its final measured value to correct the sensor data. The simulated sensor datasets and real datasets were used for simulation verification. The results show that the proposed method produces a lower mean absolute error and higher accuracy.

## 24.1 Introduction

In recent years, as a new generation of network information technology, the Internet of Things has been widely used and studied in many industries [1, 2]. The excellent operation of IoT devices requires reliable and accurate sensor data. However, the performance of IoT sensors varies, and data accuracy is easily affected by external interference [3, 4]. Therefore, it is necessary to design a correction method to improve the accuracy of the IoT sensor data, ensuring an optimal operation of devices.

Scholars at home and abroad have conducted many studies on the accuracy of sensor data from different perspectives. In [5], a predictive weighted fusion algorithm was proposed, which used the time correlation of the data for accuracy evaluation, and carried out weighted fusion according to the data credibility. Similarly, in [6], the accuracy of data fusion was improved by arranging fixed and movable low-cost sensors and used the average value of environmental measured data in previous years as compensation to fuse multiple environmental sensors' data. Based on time window and adaptive weighting, [7] proposed a data fusion method (DF-BTW) in which the

R. Lei (✉) · M. Xiang · P. Wang · R. Ma · K. Yu
Key Laboratory of Industrial Internet of Things and Networked Control, Chongqing University of Posts and Telecommunications, Chongqing 400065, China

 
L. C. Jain et al. (eds.), *Smart Communications, Intelligent Algorithms and Interactive Methods*, Smart Innovation, Systems and Technologies 257,
https://doi.org/10.1007/978-981-16-5164-9_24

time window was set to process the data by using the time correlation of data, and then the data was fused by the adaptive weighting method.

At present, most of the solutions to improve data accuracy focus on wireless sensor networks and the spatial or temporal correlation relevance of data, while neglecting the Spatial-temporal relevance on IoT terminals. Therefore, a data fusion method based on Spatial-temporal correlation (DF-BST) was proposed to correct the sensor data with errors. Each acquisition terminal used the DTW to calculate the distance between historical data, then used the exponential function to calculate the Spatial-temporal correlation degree. Finally, according to the Spatial-temporal correlation degree, weights were assigned for weighted fusion. The fusion result was used as its final measured value to correct the sensor data. This method ensures the excellent operation of IoT devices.

## 24.2 Sensor Data Processing

There were multiple data acquisition terminals in the detection area, and each terminal had multiple types of sensors. Terminals can synchronize data transmission through a time synchronization mechanism. Each terminal used the DF-BST method to calculate and obtained a fusion result as its final measured data. In this way, the original data with errors can be corrected.

### *24.2.1 Data Spatial-Temporal Correlation*

The DF-BST calculated the time data sequence's spatial distance between the acquisition terminals at each time point to reflect the sequence's similarity in temporal and spatial. Dynamic Time Warping is an algorithm used to measure data sequences' similarity, which can solve the problem of unequal data sequence lengths caused by data loss. It has been widely used in many industries [8]. In practical applications, IoT data may be lost. The dynamic time warping distance can calculate the same length under normal conditions and the sequence of different lengths when the data was lost. The distance was expressed as Eq. (24.1):

$$r\left(U_i(t), V_j(t)\right) = d\left(u_t, v_t\right) + \min\{r\left(U_i(t-1), V_j(t)\right), \\ r\left(U_i(t), V_j(t-1)\right), r\left(U_i(t-1), V_j(t-1)\right)\} \quad (t > 0) \tag{24.1}$$

In Eq. (24.1), the time series $U_i(t) = \{u_1, u_2, ..., u_t\}$ and $V_j(t) = \{v_1, v_2, ..., v_t\}$ were the sensing data acquired by the IoT terminals $i$ and $j$ during the time $t$ and the Euclidean distance between the sequence data was $d(u_t, v_t)$, the dynamic time warping distance of the sequence was $d_{\text{DTW}}(U_i(t), V_j(t)) = r\left(U_i(t), V_j(t)\right)$.

The DTW has a high time complexity and a long-running time, so a sliding window of length $t$ was set to analyze multiple terminal time data sequences [9]. The distance $d_{i,j}$ between the terminal data sequences was expressed as Eq. (24.2):

$$d_{i,j}(t) = \begin{cases} d_{\mathrm{DTW}}(U_i(t), V_j(t)) & i \neq j \\ 0 & i = j \end{cases} \quad (24.2)$$

where $U_i(t)$ and $V_j(t)$ were the time data sequence of the terminal $i$ and the terminal $j$ within the window time $t$.

Each terminal's data in the divided data acquisition area should be positively correlated in time and space. If there was data with a low degree of correlation, it should be excluded. Median Absolute Deviation is a method for detecting outliers, and it was used to detect low-correlation terminals with large Spatial-temporal distances in this method. The median operation is set as med( ). This method calculated the median of the Spatial-temporal distance, then calculated the absolute difference with each distance, and finally took the median of all absolute values as MAD. It was expressed as follows:

$$\mathrm{MAD} = \mathrm{med}\left[\left|d_{i,j} - \mathrm{med}(d_{i,j})\right|\right] \quad (j = 1, 2, \ldots, n) \quad (24.3)$$

$$Q = \mathrm{med}(d_{i,j}) + 2\mathrm{MAD} \quad (24.4)$$

The threshold $Q$ was set as the sum of the distance's median and two times MAD. The acquisition terminal that exceeded the threshold was deemed low relevance and did not participate in the subsequent fusion calculations.

The method improved the accuracy by fusing all associated terminal sensor data. The value of the distance $d_{i,j}$ would appear to be 0, which was not suitable for weight assignment. Therefore, the exponential function was used to quantify the Spatial-temporal correlation degree $s_{i,j}$ between each terminal data. The Z-score standardization method was used to process the Spatial-temporal distance $d_{i,j}$ to conform to the normal distribution [10], and the exponential function was expressed as Eq. (24.5).

$$s_{i,j} = f(d_{i,j}) = \mathrm{e}^{-d_{i,j}^2/2} \quad (24.5)$$

### 24.2.2 *Data Weighted Fusion*

Each acquisition terminal improved the accuracy of its sensing data by fusing other related terminals' sensing data. According to the Spatial-temporal correlation $s_{i,j}$ between these terminals, weights were assigned to perform weighted data fusion.

The fusion weights were allocated using the Spatial-temporal correlation obtained by Eq. (24.5), where the higher the correlation between the terminal sensor data, the greater the weight assigned.

$$\omega_{i,j}=\frac{s_{i,j}}{\sum_{k=1}^{n} s_{i,k}} \tag{24.6}$$

In the equation, $i$, $j$ and $k$ were the labels corresponding to the terminal, and the weight $\omega_{i,j}$ should satisfy:

$$\begin{cases} \sum_{j=1}^{N} \omega_{i,j}=1 \\ 0 < \omega_{i,j} \leq 1 \end{cases} \tag{24.7}$$

According to the calculated weights, weighted data fusion was performed on the sensor data of each acquisition terminal at time $t$, and the result is shown as:

$$x_i^*(t) = \sum_{j=1}^{N} x_j(t)\omega_{i,j} \tag{24.8}$$

Data $x_j(t)$ was the actual measured value of each terminal at time $t$, and $x_i^*(t)$ was the fusion result of terminal $i$ at time $t$. The fusion results were used as the final measured value to correct the original data with low accuracy.

## 24.3 Simulation Results and Analysis

Through two simulation experiments, the effective and reliability of the DF-BST method were verified. The first experiment used six sensor mathematical models to generate random datasets and used their random errors as simulated environmental interference. The second experiment used the real datasets in [11] for data fusion and compared with the experimental results.

These simulation experiments were performed in MATLAB 2017a, and used mean absolute error (MAE) and root mean square error (RMSE) to measure the accuracy of the method.

The reliability of the method was verified by setting up six mathematical models of sensors belonging to different terminals and completing experiments. The sliding window size was set to 10. The simulation data was $S_i$, and $x$ was the real value of 20. The equation was shown as follows:

$$\begin{cases} S_1 = x + n_1 \\ S_2 = x + n_2 \\ S_3 = x + n_3 \\ S_4 = x + n_4 \\ S_5 = xe^{0.0007x} + n_5 \\ S_6 = x + \sin(0.1x) + n_6 \end{cases} \tag{24.9}$$

Random noise $n_i (i = 1, 2, ..., 6)$ obeyed a normal distribution with a mean value of 0 and a variance of 0.1. Gross errors were added to $S_5$ and $S_6$ to simulate external interference.

Based on the simulation datasets, the DF-BST method, the mean method and the method in [11] were used for simulation experiments. The results were shown in Fig. 24.1 and Table 24.1.

Figure 24.1 shows that the result of the DF-BST method is closer to the real value, and the experimental result is better than other methods. The reason is that the two sensor models in this experiment increased gross errors, and the DF-BST method used the temporal and spatial correlation between the data to calculate, which can effectively reduce the errors caused by external interference.

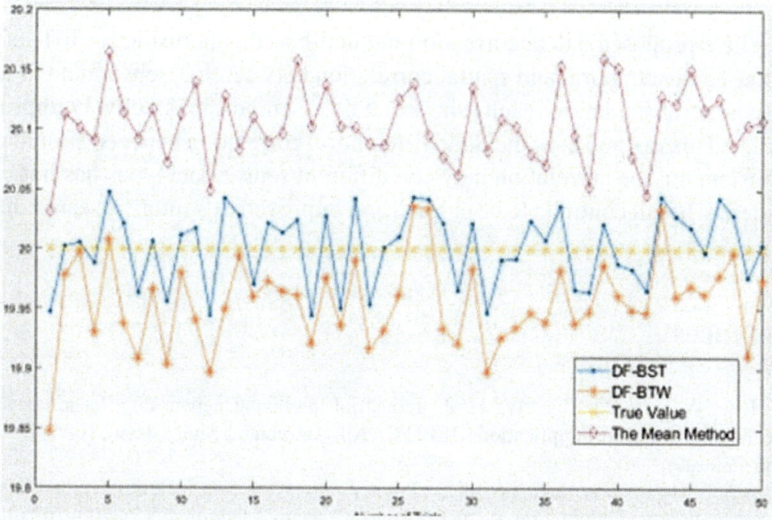


**Fig. 24.1** Results of data fusion

**Table 24.1** Comparison of data fusion results based on datasets fusion results

| | RSME | MAE |
|---|---|---|
| The mean method | 0.1145 | 0.1084 |
| DF-BTW | 0.0590 | 0.0480 |
| DF-BST | 0.0448 | 0.0362 |

**Table 24.2** Comparison of data fusion results based on real datasets

| | Fusion results | Errors |
|---|---|---|
| The mean method | 54.3967 | 0.02609 |
| DF-BTW | 52.7440 | 0.00483 |
| DF-BST | 53.0792 | 0.00179 |

As shown in Table 24.1, the DF-BST method is better than other methods in indicators such as MAE and RSME. Therefore, according to the fusion results and various indicators, it can be concluded that the DF-BST method is effective and has better performance.

The DF-BST method was used for calculation based on the datasets in [11]. The true value of the data was 53, and the comparison results were shown in Table 24.2.

As shown in Table 24.2, the error of the DF-BST method is lower. Because the DF-BTW method only considers the relevance of the sensor data in time, and the DF-BST method considers the relevance of the sensor data in time and space, the result is more accurate.

## 24.4 Conclusion

This article proposed a data correction method-based data fusion for IoT terminals, which used the temporal and spatial correlation between IoT sensor data to improve data accuracy. Simulation results show that this method can effectively improve data accuracy. This method uses the Spatial-temporal correlation between similar data for data fusion, but the correlation between different sensor data types has not yet been considered. It will continue to be studied and improved as a future research direction.

## References


1. Feiyue, W., Jun, Z., Xiao, W., et al.: Industrial intelligent networking: basic concepts, key technologies and core applications. IEEE/CAA J. Automatica Sinica **44**(9), 1606–1617 (2018). (in Chinese)
2. Lee, Y.T., Hsiao, W.H., Lin, Y.S., et al.: Privacy-preserving data analytics in cloud-based smart home with community hierarchy. IEEE Trans. Consum. Electron. **63**(2), 200–207 (2017)
3. Liqin, T., Chuan, L., Qi, Z., et al.: Reliability design and optimization analysis of IoT monitoring topology. J. Softw. **25**(08), 1625–1639 (2014). (in Chinese)
4. Yue, Y., Zuo, X., Luo, X.: Biased estimation method of multi-sensor data fusion to improve measurement reliability. IEEE/CAA J. Automatica Sinica **40**(09), 1843–1852 (2014). (in Chinese)
5. Yu, X., Fan, F., Zhou, L., et al.: Adaptive predictive weighted data fusion algorithm for wireless sensor networks. Chinese J. Sens. Actuators **30**(05), 772–776 (2017). (in Chinese)
6. Wu, H., Gao, S., Peng, H., et al.: Adaptive fuzzy C-means clustering data fusion algorithm. Comput. Eng. Appl. **55**(05), 26–35+82 (2019). (in Chinese)

7. Gressent, A., Malherbe, L., Colette, A., et al.: Data fusion for air quality mapping using low-cost sensor observations: feasibility andadded-value. Environ. Int. **143**,105965 (2020). (in Chinese)
8. Li, Z., Zhang, F., Li, K., et al.: A multivariate time series index structure supporting DTW distance. J. Softw. **25**(03), 560–575 (2014). (in Chinese)
9. Feng, Y., Ding, C., Gao, X., et al.: Time series similarity based on moving average and piecewise linear regression. Comput. Sci. **45**(S1), 110–113 (2018). (in Chinese)
10. Bi, J., Zhen, J., Wang, Y., et al.: Improved KNN indoor positioning method based on Gaussian function. Bull. Surveying Mapping (06), 9–12+35 (2017). (in Chinese)
11. Hu, W., Zhao, W.: Energy Internet data fusion method based on time window and adaptive weighting. Syst. Engi.-Theory Methodol. Appl. **25**(05), 907–913 (2016)

# Chapter 25
# Research on Multi-Sensor Data Fusion Algorithm for Monitoring of Power Distribution Station

**Peng Wang, Min Xiang, and Rujie Lei**

**Abstract** Aiming at the issue of data anomalies and excessive measurement errors caused by the limitation that monitoring nodes in power distribution stations are vulnerable to the environment, an improved data fusion algorithm based on closeness concept is proposed in this research. First, the Grubbs criterion is used to validate the validity of the data, the invalid data is removed to form a set of valid values. Then, in order to reduce the influence of measurement error on the fusion accuracy, the support degree function is introduced to improve the method of target value selection and valid value processing in the closeness-based fusion algorithm. The simulation test results show that the algorithm can effectively reduce the mean square error of the fusion result and improve the fusion accuracy.

## 25.1 Introduction

With the construction of smart grid, ubiquitous power Internet of Things projects and the development of Internet of Things technology, advanced smart sensor technology and data processing technology have been applied to monitor the environment and equipment in the distribution station, and to a certain extent, the efficiency and reliability of the monitoring process have been improved [1–4]. Hong-Xing et al. [5] designed a wireless temperature measurement system for the bus-bar temperature of the switch-gear to realize the online monitoring of the bus-bar temperature of the switch-gear in response to the issue of low safety of temperature measurement of the switch-gear bus-bar in the distribution station; [6] used multiple sensors and a data fusion method based on closeness to fuse the monitoring data of power equipment, this algorithm improved the data accuracy to a certain extent, but does not consider the measurement error caused by the distribution station environment, so the fusion accuracy will be reduced.

---

P. Wang (✉) · M. Xiang · R. Lei
Key Laboratory of Industrial Internet of Things and Networked Control, Chongqing University of Posts and Telecommunications, Chongqing 400065, China
e-mail: vitawp@163.com

 
L. C. Jain et al. (eds.), *Smart Communications, Intelligent Algorithms and Interactive Methods*, Smart Innovation, Systems and Technologies 257,
https://doi.org/10.1007/978-981-16-5164-9_25

The algorithm proposed in this research introduced the concept of comprehensive support degree and improved the data fusion algorithm based on closeness to suit the special environment of power distribution station. This algorithm uses Grubbs criterion to check the validity of data to reduce the impact of invalid data on the monitoring results caused by the environmental interference of the distribution station; The comprehensive support degree is introduced to improve the fusion algorithm based on the closeness degree, so as to reduce the interference of measurement error and improve the accuracy of the fusion result.

## 25.2 Data Fusion Algorithm

### *25.2.1 Fusion Algorithm Base on Closeness*

Assuming that the original measurement data obtained from the $n$ nodes monitoring the temperature of the power distribution room is $x_i (i = 1, 2, \ldots n)$, the closeness between the measurement data and the target value can be obtained through the closeness function based on the distance. The target value should theoretically be the actual value, but the actual value is unknown in practical applications, so the algorithm usually uses the average value of the measured data [7]. Assuming that the closeness between the original measurement value $x_i$ and the target value $\overline{x}$ is $R_i (0 \le R_i \le 1)$, the fusion algorithm bad on closeness is as follows:

$$\begin{cases} R_i = \dfrac{1}{1 + d_i} \\ d_i = |x_i - \overline{x}| \end{cases} \tag{25.1}$$

In Eq. (25.1), $R_i$ is the closeness; $d_i$ represents the distance between each original measurement value and the target value. Then use Eq. (25.2) to assign weight $\omega_i$ to each original measurement value $x_i$, and finally obtain the fusion result $Y$ through Eq. (25.3).

$$\omega_i = \frac{R_i}{\sum\limits_{i=1}^{n} R_i} \tag{25.2}$$

$$Y = \sum_{i=1}^{n} x_i \omega_i \tag{25.3}$$

### 25.2.2 Algorithm Improvement

Data preprocessing. This research used Grubbs criterion for data validity judgment. Assuming that the original data collected in the distribution room temperature was $x_i(i = 1, 2, \ldots, n)$, the Grubbs statistic $g_i$ of each data could be calculated by Eq. (25.4), among them, $x_m$ and $\sigma$ are the median and standard deviation of the data set.

$$g_i = \frac{|x_i - x_m|}{\sigma} \tag{25.4}$$

Then according to the number of data $n$ and Grubbs significance level $a$ can determine the critical threshold $g_t(n, a)$. If $g_i \geq g_t$, the data is considered invalid and should be eliminated. After preprocessing, the remaining data constituted the effective value set $y_i(i = 1, 2, \ldots, n)$, $m \leq n$.

Improvement based on comprehensive support degree. In this research, the support degree function $\sup(a, b)$ was used to express the degree of support of $b$ to $a$. The support degree function proposed by Yager must meet the following three conditions [8]:

(1) $\sup(a, b) \in [0, 1];[0, 1]$;
(2) $\sup(a, b) = \sup(b, a)$;
(3) if $|a - b| < |x - y|$, $\sup(a, b) > \sup(x, y)$.

Considered the high complexity of the support degree function proposed by Yager, in order to reduce the complexity, this research used the support degree function proposed in [9] to describe:

$$r_{ij} = \sup(y_i, y_j) = \frac{K}{1 + \beta(y_i - y_j)^4} \tag{25.5}$$

where $y_i$ and $y_j$ represent the measure values of node $i$ and node $j$; $K$ represents the magnitude of the support degree function; $\beta \geq 0$, is the attenuation factor; Then the degree of the support $r_{ij}$ and support degree matrix $D$ could be obtained:

$$D = \begin{bmatrix} r_{11} & r_{12} & \cdots & r_{1m} \\ r_{21} & r_{22} & \cdots & r_{2m} \\ \vdots & \vdots & \ddots & \vdots \\ r_{m1} & r_{m2} & \cdots & r_{mm} \end{bmatrix} \tag{25.6}$$

The *comprehensive* support degree of node $i$ is shown in Eq. (25.7):

$$s_i = \sum_{j=1}^{m} r_{ij} \tag{25.7}$$

By Eq. (25.8), we could find the value $x_h$ with support degree of $s_{\max}$ and the value $x_l$ with support degree of $s_{\min}$ in the effective value set, among them, $s_{\max}$ and $s_{\min}$ represent the highest and lowest comprehensive support degree, respectively, and the lowest comprehensive support degree indicates that the data has the largest measurement error in the effective value set, and the opposite indicates the smallest measurement error. In order to reduce the interference of the distribution station environment on the fusion result, the method of selecting the target value and the effective value *processing* in this research was improved.

First, selected the value $x_h$ with the comprehensive support degree $s_{\max}$ as the target value to realize the selection of the target value; then, in the effective value set, the value $x_l$ whose comprehensive support degree was $s_{\min}$ replaced with the value $x_h$ whose comprehensive support degree was $s_{\max}$, and the processed data $z_i$ could be obtained.

$$\begin{cases} s_{\max} = \max\{s_1, s_2, \ldots, s_m\} \\ s_{\min} = \min\{s_1, s_2, \ldots, s_m\} \\ z_i (i = 1, 2, \ldots, m),\ m \le n \end{cases} \tag{25.8}$$

Fusion result calculation. Through Eqs. (25.5)–(25.8), the target value $x_h$ and the processed *data* $z_i$ could be obtained. Replaced the original measurement data $x_i$ in Eq. (25.1) with $z_i$, and replaced the original data average value $\overline{x}$ with $x_h$, Eq. (25.9) could be obtained.

$$\begin{cases} R_i = \dfrac{1}{1 + d_i} \\ d_i = |z_i - x_h| \end{cases} \tag{25.9}$$

Then, replace the original measurement data $x_i$ in Eqs. (25.2) and (25.3) with $z_i$, and the weight $\omega_i$ and the fusion result $Y$ of the processed data $z_i$ could be obtained, respectively:

$$\omega_i = \frac{R_i}{\sum_{i=1}^{m} R_i} \tag{25.10}$$

$$Y = \sum_{i=1}^{m} z_i \omega_i \tag{25.11}$$

## 25.3 Simulation Test and Analysis

In order to validate the feasibility of the algorithm, this research used the experimental data of [10] for calculation and analysis, the specific perception data was $x_i (i = 1, 2, \ldots, 10)$, which were: 1.00, 0.99, 0.98, 0.97, 0.5, 0.65, 1.01, 1.02, 1.03, 1.5, and the actual value of this parameter was 1.00.

First, this research used Eq. (25.4) to judge the data validity. Among them, the Grubbs statistics of data with node numbers 5, 6, and 10 were 2.942, 2.706, and 2.039, respectively, which were all greater than the Grubbs critical threshold 2.036, so the data of nodes 5, 6, and 10 would be eliminated. Then used Eqs. (25.5)–(25.7) to calculate the comprehensive support degree of the remaining data. From Eq. (25.8), we could get that the comprehensive support degree of node 4 and node 7 was the lowest, which was 6.99998, and the comprehensive support degree of node 1 data was the highest at 7.00. According to the algorithm, the data of node 4 and node 7 need to be replaced with the data of node 1, and substituted the data of node 1 as the target value into Eqs. (25.9) and (25.10) to calculate the weight of each data, respectively: 0.14407, 0.14265, 0.14104, 0.14407, 0.14265, 0.14104, and 0.14407. Finally, the fusion result could be calculated by using formula (25.11):

$$
\begin{aligned}
Y = {} & 1.00 \times 0.14407 + 0.99 \times 0.14265 + 0.98 \times 0.14104 + 1.00 \times 0.14407 \\
& + 1.01 \times 0.14265 + 1.02 \times 0.14104 + 1.00 \times 0.14407 = 0.99958
\end{aligned}
$$

Among them, the result obtained by using the closeness-based fusion algorithm was 0.9679, and the result of [10] was 0.9994. It could be seen from the above results that the result of the *algorithm* in this research was closer to the actual value of 1.00, so it was feasible to use the algorithm in this research for fusion.

In order to further verify the effect of the algorithm, this research used the data in [6] for fusion. The actual value was 24 °C. Copied some data of this literature to get 12 sets of data waiting to be merged, and introduced 1 randomly generated abnormal data in groups 5~8, and 2 randomly generated abnormal data in groups 9~12 to simulate the *invalid* data generated by the environmental interference of the distribution station. The specific data is shown in Table 25.1.

The accuracy of the fusion algorithm generally uses the mean square error as the evaluation index, through the fusion calculation of the data in Table 25.1, compared the fusion results and mean square error of the improved fusion algorithm based on the closeness degree *in* this research, the closeness-based fusion algorithm in [6], and the arithmetic average method. The comparison result is shown in Fig. 25.1.

As shown in Fig. 25.1a, the fusion result of the improved closeness-based fusion algorithm is *closer* to the actual value than the closeness-based fusion algorithm and arithmetic average method, and less fluctuation. The mean square error comparison is shown in Fig. 25.1b. The improved closeness fusion algorithm has the smallest mean square error, followed by the closeness-based fusion algorithm, while the arithmetic average method has a large and unstable mean square error. The above comparison

**Table 25.1** Data to be fused

| Group number | Node 1 | Node 2 | Node 3 | Node 4 | Node 5 | Node 6 | Node 7 | Node 8 |
|---|---|---|---|---|---|---|---|---|
| 1 | 23.7 | 23.6 | 23.7 | 23.9 | 23.4 | 24.1 | 23.8 | 23.9 |
| 2 | 23.5 | 23.8 | 23.7 | 23.9 | 23.5 | 24.2 | 23.9 | 23.8 |
| 3 | 23.7 | 23.7 | 23.7 | 23.2 | 23.4 | 24.1 | 23.9 | 23.8 |
| 4 | 23.5 | 23.8 | 23.2 | 23.7 | 23.7 | 24.1 | 23.8 | 24.2 |
| 5 | 23.5 | 23.6 | 23.7 | 23.9 | 23.4 | 24.1 | 23.8 | 23.9 |
| 6 | 22.7 | 23.8 | 23.7 | 23.9 | 23.5 | 24.2 | 23.9 | 23.8 |
| 7 | 25.2 | 23.7 | 23.2 | 23.2 | 22.3 | 24.1 | 23.9 | 23.8 |
| 8 | 23.7 | 23.8 | 23.7 | 23.7 | 23.7 | 24.1 | 23.8 | 22.1 |
| 9 | 23.5 | 23.6 | 22.3 | 22.3 | 23.4 | 24.1 | 23.8 | 23.9 |
| 10 | 25.3 | 25.3 | 23.9 | 23.9 | 23.5 | 24.2 | 23.9 | 23.8 |
| 11 | 23.7 | 23.7 | 23.7 | 22.1 | 23.4 | 24.1 | 22.1 | 23.8 |
| 12 | 22.4 | 22.4 | 23.2 | 23.7 | 23.7 | 24.1 | 23.8 | 24.2 |

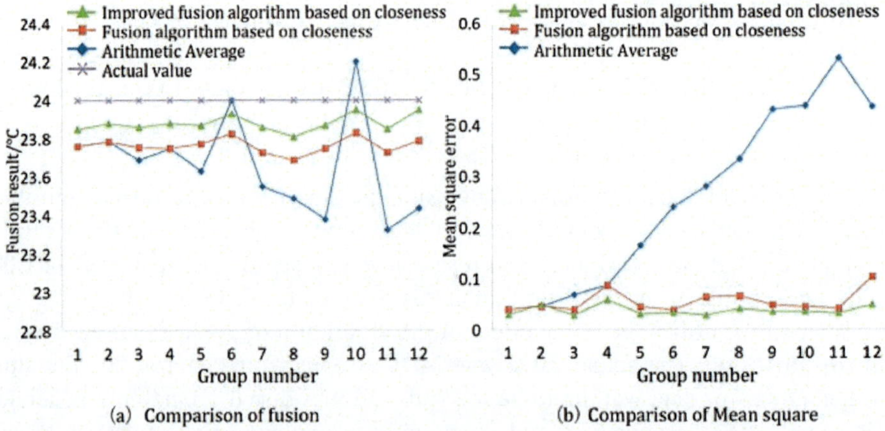


**Fig. 25.1** Comparison of fusion results and mean square error of different algorithms

results show that the algorithm in this research has obvious advantage infusion accuracy, the reason is that the data preprocessing mechanism of the algorithm in this research reduces the influence of invalid data on the fusion result. On the other hand, the improvement of the closeness fusion algorithm through the support degree function further reduces the influence of measurement errors on the fusion result and improves the accuracy of the fusion.

## 25.4 Conclusion

In order to solve the issue that the monitoring node of the distribution station is susceptible to its environmental interference, which causes data anomalies and measurement deviations, resulting in low fusion accuracy, an improved data fusion algorithm based on closeness was proposed. Firstly, the Grubbs criterion is used to identify and eliminate the invalid data generated in the monitoring process, which improves the algorithm's ability to suppress outliers. Secondly, the introduction of comprehensive support degree improves the fusion algorithm based on closeness degree, which effectively reduces the influence of measurement error on the fusion result. The simulation test results show that the fusion result of the algorithm has a low mean square error.

## References


1. Yi, W., Qi-Xin, C., Ning, Z., Cheng, F., et al: Fusion of the 5G communication and the ubiquitous electric internet of things: application analysis and research prospects. Power Syst. Technol. **43**(05), 1575–1585 (2019)
2. Gang, L., Bo, Z., Wen-Qing, Z., Yun-Peng, L., Shu-Guo, G.: Data science issues in power equipment condition assessment: challenges and prospects. Autom. Electric Power Syst. **42**(21), 10–20+177 (2018)
3. Ting, Y., Feng, Z., Ying-Jie, Z., Hai-Bo, P.: Interpretation and research prospects of ubiquitous power internet of things. Autom. Electric Power Syst. **43**(13), 9–20+53 (2019)
4. Dong-Eun, K., Hyun-Jae, L., Jin-Geun, S., Jae-Don, P.: Modified expert inference method of power substation monitoring system based on expansion of multi-sensor utilization for fire discrimination. J. Electr. Eng. Technol. **14**(3), 1385–1393 (2019)
5. Hong-Xing, F., Feng-Liang, G., Yu-Yao, X., et al: Wireless acquisition system for busbar temperature of switch cabinet based on STM32. Instrum. Tech. Sens. **4**, 77–80 (2020)
6. Shang-Cong, Z.: Research on Data Fusion Algorithm in Wireless Sensor Networks for Power Equipment Temperature Monitoring. Shanghai Dian Ji University, Shanghai (2018)
7. Yuan, Z., Jiu-Gen, Z., Jia Le, L., Xin, C.: Fuzzy nearness recommendation algorithm based on heterogeneous information network. Comput. Eng. Design **41**(02), 367–372 (2019)
8. Ronald, R.Y.: The power average operator for information fusion. In: 13th International Conference on Information Processing and Management of Uncertainty, Dortmund (2010)
9. Wen-Guang, Y., Yun-Jie, W., Jian-Min, W.: Improvement and Development of the New Model of Grey Correlation Analysis. J. Zhengzhou Univ. (Nat. Sci. Edn.) **49**(02) 25–30 (2017)
10. Jian-Shu, L., Ren-Hou, L. Hong, C.: Multi-sensor data fusion based on correlation function and least square. Control Decis. **6**, 714–716+720 (2006)

# Chapter 26
# Analysis of QR Reordering Algorithm Based on Feedback Technology Optimization

**You Xinhua, Long wenjia, and Dai Yu**

**Abstract** Feedback technology is an important tool to understand user search intentions, achieve personalized ranking, improve search engine ranking results, and increase user satisfaction with search engine results. Using feedback information to optimize the index of sorting performance and using it for resorting algorithms is a current research hotspot. This paper deeply studies the characteristics of users' feedback behavior and proposes a new evaluation indicator Quality of a Ranked list (QR). Then, on the basis of the existing classification evaluation method, the re-ranking process is improved and proposed a new and stable reordering algorithm which is introduced. The reordering algorithm proposed in this paper avoids the unreliability and noise problems of feedback documents.

## 26.1 Introduction

A typical user search method is that users use search engines as an information service form according to their needs to help users quickly and accurately find what they need in the ocean of information. The search purpose is organized into several keywords as input and sent to the search engine. The search engine searches according to the keywords and returns the search results to the user [1–4]. However, a query result often contains thousands of uneven and different contents. Often only a few of these content are in line with the user's retrieval purpose, and the accuracy of the query result set is low. Because users cannot organize their search intent into high-quality keywords, the search results are unsatisfactory. Therefore, search engines need some additional technologies to assist in mining users' query intentions and improve search results [5–7]. The use of feedback technology can be closer to the user's search intention, so that users can get the desired search results and get personalized search. The classic information retrieval model of personalized search usually consists of four elements: the logical representation of a large document collection; the logical representation of user information needs; the background description of the model;

Y. Xinhua · L. wenjia · D. Yu (✉)
Zhixing College of Hubei University, Wuhan, China
e-mail: daiyuwh@163.com

 
L. C. Jain et al. (eds.), *Smart Communications, Intelligent Algorithms and Interactive Methods*, Smart Innovation, Systems and Technologies 257,
https://doi.org/10.1007/978-981-16-5164-9_26

and the sorting algorithm [8–11]. For any user information requirement, the sorting algorithm calculates the relevance weight of each document [12]. The main function of the relevant feedback technology is to provide a new sorting algorithm, in the case of adding a known condition (user feedback information collection), recalculating the relevance weight of the query document, thereby changing the original sorting result, improve the quality of retrieval.

After getting user feedback, how to evaluate the results is an important step in related feedback technology. In the field of information retrieval, the two most classic evaluation indicators: precision and recall [13–16]. Accuracy is a measure of the ratio of found-related documents in all returned result sets, and the recall rate is a measure of the ratio of found-related documents in all related documents. However, the accuracy and the recall rate only measure the quality of the result set and do not consider the quality of the result set sorting. Literature proposed an evaluation index to measure the quality of ranking: Discounted cumulative gain (DCG) index [17–20]. Using this indicator, the order of sorting in a document collection can be evaluated.

In any system that uses user feedback, because the user gives a relevance judgment on certain documents, the system naturally prioritizes the ranking of related feedback documents. The result will definitely lead to an improvement in accuracy or recall rate, that is, "sort "Effect" improved. However, because the ranking of the documents that the user has not yet viewed may not be improved, that is, the "feedback effect" has not been improved, then such feedback results do not bring any benefit to the user. Therefore, when evaluating a feedback system, it is necessary to distinguish between "ranking effect" and "feedback effect."

Based on the existing "ranking method" and "feedback effect," this paper proposes a measurement indicator Quality of a Ranked list (QR). Then, on the basis of the existing classification evaluation method, the re-ranking process is improved and the A new stable reordering algorithm.

## 26.2 QR Index Reordering Algorithm

The sorting method of documents directly affects the "feedback effect" and ultimately affects the retrieval results of documents. Usually precision, recall, and other derivative indicators, such as MAP (Mean Average Precision) [21, 22], r-p (Recall-Precision), etc., are used to measure the quality of a sorting result. Precision refers to the proportion of documents retrieved that are related documents; recall rate refers to the proportion of how many related documents are retrieved correctly. The calculation method of MAP is: the MAP of a single query subject is the average of the similarity of each related document. The above indicators are based on the set and do not consider the impact of the ranking results. How to measure the quality of the ranking results, using the DCG indicator. The calculation formula of this indicator is Formula 26.1:

$$\mathrm{DCG}_i = \begin{cases} G_1, & \text{if}(i = 1) \\ \mathrm{DCG}_{i-1} + G_i/\log_b i, & \text{otherwise} \end{cases} \tag{26.1}$$

$\mathrm{DCG}_i$ represents the DCG value of the $i$-th document in the query result set, represents the similarity weight of the $i$-th result, and $b$ is a parameter. It can be seen from Formula 26.1 that the importance of the result scores with the lower sequence number decays according to the logarithm of the sequence number. The biggest advantage of the DCG indicator is that it has a variety of relevance levels, that is, the relevance of documents is not only represented by 0,1, but is uniformly represented in an indicator. The higher the ranking of the document, the greater the contribution to the DCG value. After the relevance of each result is manually marked, the DCG index can be used to calculate the quality of the current ranking result.

However, the DCG index can only be applied to the experimental process. In the real retrieval process, DCG cannot be used to evaluate the quality of a ranking result. This paper proposes a new indicator QR, which is proportional to the DCG indicator and can be calculated without any manual intervention. In an actual retrieval process, the QR indicator is a real and calculable indicator.

The previous reordering algorithms mainly focused on the relevance of a candidate document to user interests. Score each candidate document according to certain indicators and sort all candidate documents according to the score. The focus is to examine the quality of a document and examine the similarity between a single document and the user's intention. This article solves the reordering problem from another perspective: enlarge the granularity of the object to be examined and examine the characteristics that a document sequence (a ranking result) should have when the overall relevance of it is better. The measure of the quality of a ranking result can be calculated from the characteristics of the ranking result.

A retrieval intention of a given user q and an initial document collection $\{d_1, d_2, \ldots, d_n\}$, let the ideal sorting result is $\langle d_1, \ldots d_i, d_{i+1} \ldots, d_n \rangle$. In this ideal ranking, all documents are arranged in descending order of relevance to the user's search intent; $\langle d_1, \ldots d_i \rangle$ are related documents (that is $\langle d_1, \ldots d_i \rangle$, their relevance is greater than the user's minimum expected threshold for retrieval results), and $\langle d_{i+1} \ldots, d_n \rangle$ are irrelevant documents. By studying the ideal result set, it can be assumed that the ideal result set should have the following characteristics:

Sort: Documents in $\{d_1, d_2, \ldots, d_n\}$ are sorted in descending order of relevance. Since the similarity is only the result of guessing, this is the most basic and easiest condition to satisfy, and it is also an impossible condition. The QR measurement index proposed in this article only needs to complete a descending order at the end.

Radius: The average distance from each document to the centroid is radius in $\langle d_1, \ldots d_i \rangle$.

$$\text{radius} = \frac{1}{\text{relnum}} \sum_{j=1}^{\text{relnum}} \text{dist}\left(d_j, d_{\text{center}}\right) \tag{26.2}$$

where radius represents the radius of the first $i$ documents; relnum represents the number of related documents $i$; the dist function is a distance function, used to calculate the distance between the document $d_j$ and the centroid $d_{\text{center}}$ of the previous i documents. In the ideal result sequence, all related documents are ranked first, and they are more similar to each other (greater than the similarity to unrelated documents). Therefore, in a ranking result with better quality, if more related documents are ranked higher, the radius of the first $i$ documents is also smaller.

Document ratio: In the document collection $\langle d_{i+1} \ldots, d_n \rangle$, that is sorted lower, count how many documents may be similar documents. Calculated by Formula (26.3), where ratio represents the document ratio, $d_{\text{center}}$ represents the centroid of the similar document $\langle d_1, \ldots d_i \rangle$ set, radius represents the radius of the similar document set, and *irelset* represents the irrelevant document set, that is $\langle d_{i+1} \ldots, d_n \rangle$.

Assuming that $d_j$ is sufficiently similar to similar documents (that is, the distance between $d_j$ and the centroid $d_{\text{center}}$ of the similar document set is less than the radius of the similar document set), it is considered that $d_j$ may be a similar document that has been missed. The more similar documents that are missed (that is, the higher the proportion of documents), the worse the quality of the resulting sequence.

Combine the above three features to construct QR indicator:

$$\text{QR}(D_n) = \frac{\alpha}{\text{radius}} + \frac{\beta}{\text{ratio}} \tag{26.3}$$

where $D_n$ represents the sorting quality of a result sequence $\{d_1, d_2, \ldots, d_n\}$, radius represents the radius of related documents, ratio represents the possible proportion of related documents, and is an empirical adjustment parameter used to adjust the influence of radius and ratio on the final QR value.

Usually $n$ is 200, which means that the first 200 search results are reordered. Although for a query, search engines may return thousands of search results; on the one hand, users often only care about the sorting results of the first few pages; on the other hand, because the re-sorting process must be completed in real time; therefore, in a user feedback, in the system, the processed document collection is usually only the initial search results of the first few pages. In such a document collection, it can be seen that the value range of the radius and the value range of the ratio document ratio are quite different. In order to balance the influence of these two factors on the QR indicator, an empirical adjustment function log is added, and the QR calculation formula is:

$$\text{QR}(D_n) = \log\left(\frac{1}{\text{radius}}\right) + \frac{1}{\text{ratio}} \tag{26.4}$$

The QR indicator is directly proportional to the DCG indicator, which means that the QR indicator can measure the quality of a ranking result.

## 26.3 Improved Reordering Algorithm Using QR Indicator

The usual user feedback algorithms are mainly divided into two types: user feedback algorithms based on optimized queries and user feedback algorithms based on document comparison. There are three main reordering algorithms for user feedback:

(1) IDE-Regular algorithm [12]. A classic query expansion and weight adjustment algorithm in the vector space model (VSM).

$$\text{Ide-Regular: } \vec{q}_m = \alpha\vec{q} + \beta \sum_{\forall d_j \in D_r} \vec{d}_j - \gamma \sum_{\forall d_j \in D_n} \vec{d}_j \tag{26.5}$$

IDE-Regular algorithm is a basic algorithm for user feedback based on optimized queries, where $q_m$ is new adjusted query, $q$ is original query, $D_r$ is collection of relevant user feedback documents, and $D_n$ is the user's negative feedback document collection, $\beta = 0.75$, $\gamma = 0.25$, $\alpha = 0.25$.

(2) Linear comparison algorithm. Given n feedback documents $\{f_1, f_2, \ldots, f_n\}$, the relevance of candidate document $d_i$ is obtained by linear addition of $n$ (wherein represents the similarity between document $d_i$ and feedback document $f_i$). Use it as a basic algorithm for user feedback based on document comparison.

$$\text{relevance of}(d_i | f_1, f_2, \ldots f_n) = \sum_{j=1}^{n} w_{f_j, d_i} \tag{26.6}$$

(3) QR weighting algorithm. This algorithm is proposed in combination with the QR indicator. For each feedback document, the quality of the current ranking result that is $\mathrm{QR}_{\mathrm{current}}$ value when it is clicked is calculated; this is used to measure the reliability of different feedback documents. When the QR value is high, the predicted relevance of the document is high, and it should have a high influence on the reordering process, so the weight is relatively large.

$$\text{relevance of}(d_i | f_1, f_2, \ldots f_n) = \sum_{j=1}^{n} \mathrm{QR}_{\mathrm{current}} w_{f_j, d_i} \tag{26.7}$$

On the basis of obtaining QR, the sorting algorithm is further improved. In most cases, the user's click data is directly used as the relevant feedback information. However, a result document being clicked by the user does not mean that the result document must be a relevant document. There are many reasons that affect the user's click behavior, and the quality of the current sorting result is one of the main reasons: when the current sorting result is of good quality, the user's click document can well reflect the user's needs, when the quality of the current ranking results is not as good as expected, the user will take the second place and click on some less relevant documents, that is, "pull generals from short ones." This phenomenon is

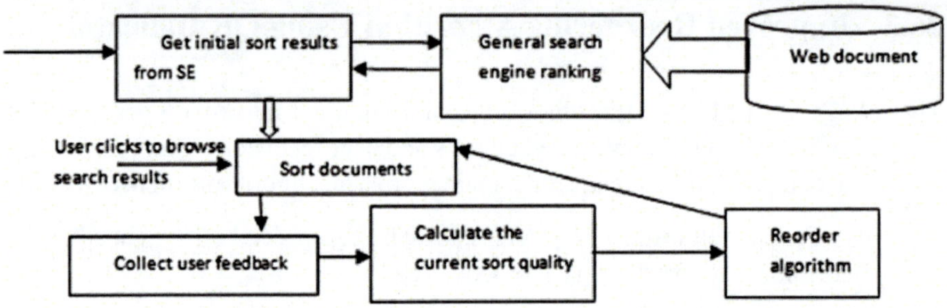


**Fig. 26.1** Reordering algorithm flow

called quality deviation, that is, the quality of the current ranking results affects the quality of the documents that users click on. Therefore, the processing of user click data should consider the quality of the current ranking results.

The current ranking result is of better quality, that is, when the QR value is high, and the feedback document clicked by the current user may be more relevant; the feedback document should have a greater influence on the re-ranking process. Therefore, the core idea of the new reordering algorithm is to use the QR value to weight different feedback documents to reflect the different importance of each feedback document. The flow of the algorithm and its relationship with general search engines is shown in Fig. 26.1.

## 26.4 Conclusion

Based on the in-depth study of user feedback behavior characteristics, in order to avoid the unreliability and noise problems of feedback documents, this paper introduces classification technology to improve the reordering process and improve the efficiency of the classification algorithm. After carefully evaluating and examining the existing evaluation methods for classification, a new stable evaluation method QR algorithm is proposed to improve the quality of the ranking results. On the basis of the QR indicator, the relevance of different feedback documents can be calculated by the quality of the current ranking results, which is the idea of the QR weighted ranking algorithm. In this way, the feedback documents with higher relevance play a more important role in the reordering process and ultimately effectively improve the quality of the ranking results.

**Acknowledgements** This work described in this paper was fully supported by a grant from the Guiding Project of Science and Technology research plan of Hubei Provincial Department of Education (No: B2020329).

## References


1. Baeza-Yates, R., Ribeiro-Neto, B.: Modern Information Retrieval. Pearson Education (2014)
2. Ruthven, I.: Re-examining the potential effectiveness of interactive query expansion. SIGIR'09 Proceedings of the 32th Annual International ACM SIGIR Conference on Research and Development in Information Retrieval, pp. 213–220, Boston, USA (2012)
3. Dumais, S., Cutrell, E., Cadiz, J.J., Jancke, G., Sarin, R., Robbins, D.C.: Stuff I've Seen: A system for personal information retrieval andre-use. Proceedings of SIGIR (2016)
4. Morita, M., Shinoda, Y.: Information filtering based on user behavior analysis and best match text retrieval. ACM Conference on Research and Development in Information Retrieval (SIGIR) **49**(2), 28–35 (2015)
5. Joachims, T., Granka, L., Pan, B.: Accurately interpreting clickthrough data as implicit fe. SIGIR 2005 **51**(1), 4–11 (2016)
6. Lundquist, C., Grossman, D.A., Frieder, O.: Improving relevance feedback in the vector space model. Proceedings of the CIKM Conference (CIKM97), LasVegas Nevada, USA (2010)
7. Rocchio, J.J.: Relevance feedback in information retrieval. The Smart Retrieval System-Experiments in Automatic Document Processing. G.Salton, Prentice Hall Inc., Englewood Cliffs, NJ (2016)
8. Ide, E.: New experiments in relevance feedback, pp. 337–354. The Smart Retrieval System. G. Salton, Prentice Hall (2012)
9. Lee, J.H.: Combining multiple evidence from different relevance feedback methods. Inf. Process. Manag. **34**(6), 681–691 (2017)
10. Yu, C.T., Buckley, C.: A generalized term dependence model in information retrieval. Inf. Technol. Res. Develop. **2**(4), 129–154 (2016)
11. Ponte, J.M., Croft, W.B.: A language modeling approach to information retrieval. SIGIR'98 In 21st ACM SIGIR Conference on Research and Development in Information Retrieval (SIGIR'98) (2010)
12. Hiemstra, D., Kraajj, W.: Twenty-one at TREC-7:ad-hoc and cross-language track. Proceedings of Seventh Text Retrieval Conference (TREC-7) (2015)
13. Kurland, O., Lee, L., et al.: Better than the real thing: iterative pseudo-query processing using cluster-based language models. IGIR'05 (2005)
14. Kelly, D., Teevan, J.: Implicit Feedback for Inferring User Preference: A Bibliography. sigir forum (2003)
15. Baeza-Yates, R., Ribeiro-Neto, B.: Modern Information Retrieval. Pearson Education (2004)
16. Ruthven, I.: Re-examining the potential effectiveness of interactive query expansion. SIGIR'09 Proceedings of the 32th Annual International ACM SIGIR Conference on Research and Development in Information Retrieval, Boston, USA (2009)
17. Dumais, S., Cutrell, E., et al.: Stuff I've Seen: A system for personal information retrieval and re-use. Proceedings of SIGIR. (2006)
18. Morita, M., Shinoda, Y.: Information filtering based on user behavior analysis and best match text retrieval. ACM Conference on Research and Development in Information Retrieval (SIGIR) (1994)
19. Joachims, T., Granka, L., et al.: Accurately interpreting clickthrough data as implicit fe. SIGIR (2005)
20. Voorhees, E.: Evaluation by highly relevant documents. SIGIR'2001 Proceedings of the 24th ACM-SIGIR Conference on Research and Development in Information Retrieval, New York (2001)
21. Buckley, C., Voorhees, E.M.: Retrieval System Evaluation. TREC: Experiment and Evaluation in Information Retrieval. MIT Press (2005)
22. Jarvelin, K., Kekalainen, J.: IR evaluation methods for retrieving highly relevant documents. Proceedings of the 23rd Annual International ACM SIGIR Conference on Research and Development in Information Retrieval (2000)

# Chapter 27
# Example Verification of a Safety Evaluation Model for Road Passenger Transportation

**Fujia Liu, Chaozhou Chen, and Guoliang Dong**

**Abstract** The development of network technology makes it possible for transportation companies to quickly collect and integrate information during transportation. The information extracted from different information management systems can be used for risk assessment and safety warning of transportation vehicles. Using the safety evaluation model can evaluate the risks in the transportation process in advance and prevent accidents. The quality of the evaluation model will affect the objectivity of the conclusion. This article introduces the evaluation indicators for safety evaluation. Aiming at the problem that the evaluation conclusions in the traditional evaluation model do not match the actual situation, an improved model method is proposed. The two actual cases were analyzed and compared with the evaluation model before and after the improvement. It was found that the improved evaluation model solved the problems of the traditional evaluation model, and the evaluation conclusions were consistent with the actual situation.

## 27.1 Introduction

Once an accident occurs on a road passenger transport vehicle, it will cause great harm to the occupants and society. Evaluation of the safety status of the transportation process before the departure of the vehicle can predict risks in advance. Network management technology is widely used in the transportation process, including driver information management, vehicle technical status and maintenance information management, wireless and network dynamic monitoring, and many other aspects. The information in the information system and wireless transmission is input into the risk evaluation model, and the risk of the transportation process can be evaluated. The quality of the evaluation model determines the quality of the safety evaluation. Inappropriate evaluation models will lead to large deviations in evaluation results, which cannot meet the safety management requirements of transportation companies. Based on the study of the classic evaluation model, an improved transportation

F. Liu (✉) · C. Chen · G. Dong
Research Institute of Highway Ministry of Transport, Beijing 100088, China
e-mail: fj.liu@rioh.cn

 
L. C. Jain et al. (eds.), *Smart Communications, Intelligent Algorithms and Interactive Methods*, Smart Innovation, Systems and Technologies 257,
https://doi.org/10.1007/978-981-16-5164-9_27

process safety evaluation model is proposed. Using the evaluation model before and after the improvement to analyze and compare two actual cases, it is found that the improved evaluation model solves the problems in the traditional evaluation model, and the evaluation conclusion is consistent with the actual situation.

## 27.2 Evaluation Index

The safety of transportation involves many factors such as drivers, vehicles, roads and environment, management, and so on. Accident statistics show that 85% of accidents are caused by the fault of the driver. Speeding and driving in violation of safe operating regulations are the main manifestations of driver faults. Drivers' weak safety awareness leads to serious violations which are one of the main reasons for road traffic accidents [1–3]. The driver's safety awareness determines the standardization of his driving behavior and safety awareness affects risk perception and driving skills [4]. The relationship between driving experience and safety is not linear [5–7]. Drivers who have been driving for less than 5 years have the greatest safety risk. Drivers younger than 25 years old have the greatest risk of driving safety. Drivers under 30 and over 57 are at higher risk. Drivers in the 35–40 and 49–54 age groups have lower safety risks. For those who have been driving for 5–10 years, the safety risks gradually increase. If the driving experience is more than 10 years, the risk gradually decreases. The number of accidents caused by mechanical failure is relatively small [7]. The key vehicle components that affect vehicle safety mainly include: brakes, steering, tires, seat belts, etc. Road alignments such as long slopes, steep slopes, and roads near rivers will have a greater impact on driving safety. In particular, conditions such as slippery roads, water accumulation, and icing caused by rain and snow weather are more likely to cause accidents under the above road conditions. The safety management of vehicles and drivers by transportation companies can improve the safety awareness of drivers and ensure that the technical conditions of the vehicles are in good condition. The regular safety education of drivers by transportation companies will significantly reduce accidents [8].

## 27.3 Evaluation Model

### 27.3.1 *Evaluation Method*

The fuzzy comprehensive evaluation method used for the safety evaluation of the transportation process integrates the advantages of the analytic hierarchy process and has the advantages of simple calculation and reasonable evaluation results [9]. In the evaluation, it is necessary to calculate the membership degree subset for each evaluation index. For qualitative and quantitative indicators, different methods should be

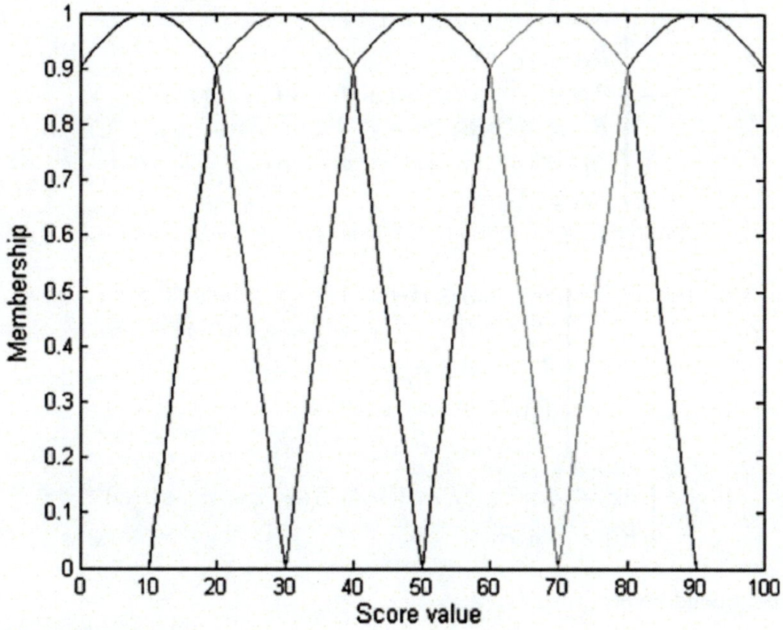


**Fig. 27.1** Membership function based on the Cauchy distribution

used [10]. Indicators such as driving style, maintenance conditions, weather conditions, and road conditions are all qualitative indicators, and an expert method can be used to construct a subset of membership degrees. For quantifiable indicators, the evaluation score of the indicator can be obtained according to the scoring rules, and then, the membership function is used to convert the evaluation score into a subset of membership. Driving age, age, vehicle age, vehicle mileage, etc., are quantitative indicators. The assignment method is used to determine the membership function in the safety evaluation of the transportation process. Based on the Cauchy distribution in the safety evaluation of transportation process, the membership function curve of each evaluation level is determined, as shown in Fig. 27.1.

## 27.3.2 Traditional Evaluation Model

The evaluation index is divided into three layers. The evaluation model requires that the values of the indicators in the sub-layer should be mutually exclusive, and the sum of the weight values of all the indicators in the sub-layer should be 1. The vectors $A$, $R$, and $B$ are used to represent the weight value, evaluation matrix, and membership degree of each layer. The model of each layer index is as follows:

The third layer:

$$\begin{cases} A_{11} = (a_{111}, a_{112}, a_{113}) \\ B_{11} = A_{11}.R_{11} \\ A_{12} = (a_{121}, a_{122}, a_{123}, a_{124}, a_{125}, a_{126}) \\ B_{12} = A_{12}.R_{12} \\ A_{13} = (a_{131}, a_{132}, a_{113}, a_{134}, a_{135}, a_{136}, a_{137}) \\ B_{13} = A_{13}.R_{13} \end{cases} \tag{27.1}$$

$$\begin{cases} A_{21} = (a_{211}, a_{212}) \\ B_{21} = A_{21}.R_{21} \\ A_{22} = (a_{221}, a_{222}, a_{223}, a_{224}, a_{225}) \\ B_{22} = A_{22}.R_{22} \\ A_{23} = (a_{231}, a_{232}, a_{233}) \\ B_{23} = A_{23}.R_{23} \end{cases} \tag{27.2}$$

$$\begin{cases} A_{31} = (a_{311}, a_{312}, a_{313}, a_{314}, a_{315}, a_{316}, a_{317}, a_{318}) \\ B_{31} = A_{31}.R_{31} \end{cases} \tag{27.3}$$

The second layer:

$$\begin{cases} A_1 = (a_{11}, a_{12}, a_{13}) \\ R_1 = (B_{11}, B_{12}, B_{13})^{\mathrm{T}} \\ B_1 = A_1.R_1 \end{cases} \tag{27.4}$$

$$\begin{cases} A_2 = (a_{21}, a_{22}, a_{23}) \\ R_2 = (B_{21}, B_{22}, B_{23})^{\mathrm{T}} \\ B_2 = A_2.R_2 \end{cases} \tag{27.5}$$

$$\begin{cases} A_3 = (a_{31}, a_{32}, a_{33}, a_{34}) \\ R_3 = (B_{31}, R_{32}, R_{33}, R_{34})^{\mathrm{T}} \\ B_3 = A_3.R_3 \end{cases} \tag{27.6}$$

$$\begin{cases} A_4 = (a_{41}, a_{42}, a_{43}, a_{44}, a_{45}) \\ R_4 = \left(R_{41}, R_{42,}, R_{43}, R_{44}, R_{45}\right)^{\mathrm{T}} \\ B_4 = A_4.R_4 \end{cases} \tag{27.7}$$

The first layer:

$$\begin{cases} A = (a_1, a_2, a_3, a_4) \\ R = (B_1, B_2, B_3, B_4)^{\mathrm{T}} \\ B = A.R \end{cases} \tag{27.8}$$

Evaluation result $k$:

$$k = B.C = (b_1, b_2, \ldots, b_n) \cdot \begin{bmatrix} c_1 \\ c_2 \\ \vdots \\ c_n \end{bmatrix} = \sum_{i=1}^{m} b_j \cdot c_j \tag{27.9}$$

According to the section to which the evaluation score belongs, the risk level of this transportation process can be determined.

### 27.3.3 Improved Evaluation Model

When the traditional evaluation model is used to evaluate the safety of the actual transportation process, there are often problems that the evaluation results are inconsistent with the actual situation. After analyzing the calculation process of the evaluation model, the main reasons are found.

(1) The indicators in each sub-layer of the conventional evaluation model are generally mutually exclusive. However, a single option cannot objectively express the actual situation. Therefore, for some special sub-layers, it should be allowed to check multiple items under the sub-layer.
(2) The traditional evaluation model stipulates that the sum of the weights of all evaluation indicators under the sub-layer should be 1. In the process of merging and superimposing indicator items, the degree of influence of high-weight indicators will be weakened. Especially when multiple main indicators are concentrated under the same sub-layer, the degree of weakening of the importance of these indicators is even more serious. In the improved model, if the indicators under the sub-layer are not mutually exclusive, the sum of the weights of all indicators under the sub-layer no longer has to be 1.

## 27.4 Comparison Calculation of Two Examples

In order to verify the evaluation effect of the improved evaluation model, two actual cases were evaluated to compare the risk of the evaluation model before and after the improvement. One of the cases was a traffic accident, and the other was a safe driving record of a company.

### 27.4.1 Example 1: A Traffic Accident

In an accident in 2018, a bus collided with a small bus. Many passengers on the bus were thrown out of the bus, resulting in 8 deaths. The bus was 5.5 years old, the tires were severely worn, and some of the seat belts were damaged. The bus driver is 33 years old, 14 years of driving experience, and 13 illegal records. The accident section was a straight section, and the road surface was smooth. It was light rain when the accident happened. The road is wet. No water, no fog. The passenger transport company did not perform well in monitoring the vehicles.

The driver's "physical fitness" is calculated as "average," with an evaluation score of 75 points. The driving experience is 14 years, and the evaluation score is 72 points. The age is 33 years old, and the evaluation score is 74 points. The vehicle age is 5.5 years, and the score is 77.5 points.

#### 27.4.1.1 Traditional Method for Evaluation

The following are the calculation steps of the sub-layer of "driver."

$$R_{11} = \begin{bmatrix} 0.0190.2450.4900.2450.000 \\ 0.0290.3320.4540.1840.000 \\ 0.0220.2710.4850.2220.000 \end{bmatrix}$$

$$\begin{aligned} B_{11} &= A_{11}.R_{11} = (0.0240.2860.4760.2160.000) \\ B_{12} &= A_{12}.R_{12} = (0.0520.3230.4110.1940.000) \\ B_{13} &= (0.0000.0500.0910.2020.657) \\ B_{21} &= A_{21}.R_{21} = (0.0270.2900.4150.2230.045) \\ B_{22} &= (0.0050.0710.1680.3830.373) \\ B_{23} &= A_{23}.R_{23} = (0.0130.1750.3560.2940.162) \\ B_{31} &= (0.0190.2450.4900.2450.000) \end{aligned}$$

The evaluation matrix of the driver index is:

$$R_1 = \begin{bmatrix} 0.0240.2860.4760.2160.000 \\ 0.0520.3230.4110.1940.000 \\ 0.0000.0500.0910.2020.657 \end{bmatrix}$$

The membership value of the driver index is:

$$B_1 = A_1.R_1 = (0.0220.1880.2740.2020.308)$$

Using the same method, the membership degrees of the other indexes are:

$$\begin{aligned} B_2 &= A_2.R_2 = (0.0180.1660.2960.3080.210) \\ B_3 &= A_3.R_3 = (0.0130.1650.3490.3200.155) \\ B_4 &= A_4.R_4 = (0.1600.1400.1360.2750.289) \end{aligned}$$

The evaluation matrix of the first level indicators is:

$$R = \begin{bmatrix} B_1 \\ B_2 \\ B_3 \\ B_4 \end{bmatrix} = \begin{bmatrix} 0.0220.1830.2680.2060.316 \\ 0.0190.1690.3000.3060.204 \\ 0.0130.1650.3490.3200.155 \\ 0.1600.1400.1360.2750.289 \end{bmatrix}$$

The membership degree of the first level index is:

$$B = A.R = (0.0520.1690.2560.2590.261)$$

The risk evaluation score of this transportation process calculated by the traditional evaluation method is:

$$k = B.C = \sum_{i=1}^{mn} b_j \cdot c_j = 80.1$$

Therefore, the traditional evaluation method is used to evaluate the safety risk of the transportation process before the accident with a score of 80.1, which lies in the critical zone of "higher" risk and "normal" risk. The evaluation conclusion contradicts the actual situation.

#### 27.4.1.2 Improved Method for Evaluation

Using an improved evaluation method for calculation, the safety risk evaluation score of the transportation process before the accident is 95.3 points, and the risk level is "very high." The evaluation conclusion is consistent with the actual situation.

### *27.4.2 Example 2: A Safe Transportation Process*

In December 2019, a bus safely completed a long-distance transportation task. The driver is 48 years old and has been driving for 24 years. He has violated the law for speeding below 10% once in the past year. The vehicle is 7 years old. The key components are normal and maintained in good condition. The road passed by is normal and straight. The transportation has a good management of vehicles and drivers.

The driver is 48 years old, and the evaluation score is 60 points. The driving experience is 24 years, and the evaluation score is 60 points. The driver's "physical fitness" is "good," and the evaluation score is 65 points. The driver is "familiar" with the driving route and scores 65 points. The driving style is "cautious" and scores 60 points. The driving technique is "very good," with a score of 55 points. The driver's condition before departure was "very good," with a score of 55 points. The driver violated the law once and scored 75 points. The vehicle is 7 years old and scores 85 points.

The risk evaluation score of this transportation process calculated by the traditional evaluation method is 68.3 points, which lies in the critical zone of "lower" risk and "normal" risk. The evaluation conclusion is consistent with the actual situation.

The risk evaluation score of this transportation process calculated by the improved evaluation method is 67.0 points, and the risk level is "low." The evaluation conclusion is consistent with the actual.

### *27.4.3 Comparative Analysis*

For the above two cases, the evaluation models before and after the improvement were used to calculate the individual evaluation values of driver, vehicle, road and environment, and management factors. The evaluation results are shown in Tables 27.1 and 27.2.

It can be seen from Table 27.2 that the overall evaluation results of accident cases using the model before and after the improvement are quite different (19.3%). The calculation results of the two models of the "vehicle" indicator and the "management factor" indicator differ greatly, which are 49.4% and 15.4%, respectively. The "vehicle" indicator has cancelled the requirement that the sum of the weight indicators within the layer is 1, and the single evaluation result is 119.4, breaking the limit of 100. Compared with the result calculated using the traditional evaluation

**Table 27.1** Evaluation results of an accident case

| Items | | Improved | Traditional | Difference (%) |
|---|---|---|---|---|
| Overall evaluation | | 95.3 | 79.9 | 19.3 |
| Individual evaluation | Driver | 84.9 | 80.4 | 5.5 |
| | Vehicle | 119.4 | 79.9 | 49.4 |
| | Road and environment | 82.3 | 79.5 | 3.6 |
| | Management | 91.2 | 79.0 | 15.4 |

**Table 27.2** Evaluation results of a safe transportation

| Items | | Improved | Traditional | Difference (%) |
|---|---|---|---|---|
| Overall evaluation | | 67.0 | 68.2 | −1.7 |
| Individual evaluation | Driver | 67.2 | 69.0 | −2.5 |
| | Vehicle | 69.2 | 67.9 | 2.00 |
| | Road and environment | 78.5 | 77.0 | 2.0 |
| | Management | 59.1 | 63.3 | −6.7 |

method (79.9 points), the calculated result using the improved model (95.3 points) objectively reflects the degree of risk in this transportation process.

It can be seen from Table 27.2 that the overall evaluation results of accident cases using the model before and after the improvement are not much different (1.7%). The calculation results of the "management factor" differ by 6.7%. The calculation results of other indicators differ by less than 2.5%. The evaluation model before and after the improvement has little difference for the safe transportation process.

## 27.5 Conclusion

The extensive application of information technology in the transportation process has improved the efficiency of data collection and can provide a large amount of useful information for the risk assessment of the transportation process. The analysis of these multi-source data can realize the risk assessment of the transportation process. This paper introduces the evaluation indicators for the safety evaluation of road passenger transportation. The reasons for the inconsistency between the traditional evaluation model and the actual situation are analyzed, and an improved evaluation model program is proposed. The evaluation model before and after the improvement is applied to evaluate two actual cases. After comparing the evaluation results, it is found that the traditional evaluation model can basically reflect the status of the safe transportation process, but there is a big deviation in the evaluation of the dangerous transportation. The improved evaluation model for safe transportation process and dangerous transportation process evaluation results is consistent with the actual. The comparative results of case analysis show that the improved evaluation model is effective.

**Acknowledgements** This paper was funded by the 2020 Basic Scientific Research Business Expense Project of Research Institute of Highway Ministry of Transport, "Research on the Technical Management Methods of Road Transport Vehicles Based on Big Data" (2020-9034).

## References

1. Wu, B.J., Sun, Q., et al.: Analysis of key influencing factors of automobile transportation safety management. Logistics Eng. Manag. **040**(001), 158–160 (2018)
2. Zhang, Q.: Research on driver safety awareness assessment, Ph.D. Thesis, Chang'an University (2010)
3. Gao, L.N.: Research on evaluation method of driver's safety awareness simulation driving, Ph.D. Thesis, Kunming University of Science and Technology (2015)
4. Tao, J.: In-depth analysis of drivers of commercial vehicle traffic accidents. Shanxi Transp. Sci. Technol. **253**(04), 155–158 (2018)
5. Yuan, L., Yuan, H.W., et al.: Analysis of the influence of driver's age on traffic risk perception ability. Soc. Psychol. Sci. **6**, 91–94 (2013)

6. Shi, N.N.: Analysis of the frequency distribution of driver accidents and its influencing factors. Beijing Jiaotong University, Ph.D.thesis (2018)
7. Zhu, S.S.: Research on the traffic safety risk of operating buses based on the accident tree analysis, Ph.D. Thesis, Chang'an University (2016)
8. Wang, X.S., Luo, Y., et al.: Traffic safety education for operating drivers based on risky driving behavior analysis. Automob. Safety **1**, 62–66 (2016)
9. Wu, Z.J., Fan, W.J.: Research on risk assessment system of road transportation of dangerous goods. Highw. Transp. Sci. Technol. **2**(12), 6–11 (2015)
10. Xiao, W.: Research on transportation safety evaluation of automobile units, Ph.D. Thesis, Nanjing University of Science and Technology (2009)

# Chapter 28
# Development of Broadband Sensors Based on Fiber Optic Gyroscope

**Bo Yang, Yong Li, Yafei Zhao, Peiyong Zhang, Li Sun, Xin Hui, and Xiaona Zhou**

**Abstract** Due to the observations of out small disturbance in spacecraft, a promising new field of research in inertial attitude sensor has been developed. However, exploration of new spacecraft under high-frequency oscillation has been hampered by the lack of a portable, reliable and highly sensitive broadband sensor. In this paper, the laboratory tests of a broadband inertial attitude sensor named micro-vibration measurement sensor (MMS) in spacecraft based on interferometric fiber optic gyroscope (IFOG) is firstly reported. The noise level by operating range diagrams and Allan deviation is estimated and tests of the sensors, which are demonstrated the reliability of MMS in filed installations. This new technology presents a wide range of field applications in spacecraft, earthquake engineering and seismology.

## 28.1 Introduction

For last few years, inertial attitude sensor (INS) has been looking for a reliable and highly broadband sensitive sensor for application in spacecraft which need high-frequency attitude information in the field of spacecraft attitude control with high stability [1–3]. In theory, interferometric fiber optic gyroscope (IFOG) has a high bandwidth due to its principle and will play an important role in obtaining attitude information. Therefore, well-known institutions at home and abroad have invested a lot of energy to develop a new production in this field [4–6].

For example, three SRS-2000 fiber optic gyroscopes from Optolink is the cutting edge in inertial rotational velocity measurements with the Bias drift of 0.003°/h and the random walk of $0.0002°/\sqrt{h}$, which is shown in Fig. 28.1 [7]. This device has a bandwidth of 50 Hz, which can also be triggered at up to 2 kHz. This kind of device can be used to estimate the attitude in the inertial reference frame. And it has been first launched in Fall 2016 at National Aeronautics and Space Administration (NASA) Goddard Space Flight Center for the purpose of studying galactic clustered star formation by providing spatially resolved spectroscopy of nearby star clusters.

B. Yang (✉) · Y. Li · Y. Zhao · P. Zhang · L. Sun · X. Hui · X. Zhou
Beijing Institute of Control Engineering, Beijing 100094, China
e-mail: yangbo3504@126.com

 
L. C. Jain et al. (eds.), *Smart Communications, Intelligent Algorithms and Interactive Methods*, Smart Innovation, Systems and Technologies 257,
https://doi.org/10.1007/978-981-16-5164-9_28

**Fig. 28.1** SRS-2000 and Astrix 200 fiber optic gyroscope

(a)SRS-2000 fiber optic gyroscope

(b)Astrix 200 fiber optic gyroscope

The Astrix200 is the highest precision fiber optic gyroscope for a mission led by the American, French, Canadian, and United Kingdom Space Agencies (UKSA), NASA, Centre National d'Etudes Spatiales (CNES), which launched in 2020 to study Earth's surface water, which shown in Fig. 28.1. This device has been flying on Pleades Satellites and the flight experience of 13 satellites effectively verified the reliability and performance of the products.

Nowadays, with the increasing demand of high resolution and high-reliability satellite platform for high-speed attitude measurement, it is very important to achieve stable control of satellite, while obtaining as much attitude information as possible. In this paper, a broadband micro-vibration measurement sensor (MMS) which can be used in satellite control is introduced.

## 28.2 Instrument and Method

The technology reported in this work is based on a closed-loop interferometric fiber optic gyroscope, which means that its operating point is kept at the point of the highest sensitivity by an internal feedback loop. And the diagram of the proposed MMS shown in Fig. 28.2, which consists of optical circuit and circuit. In our scheme, after being amplified and filtered, the weak signal received by the detector is converted into digital signal by analog-to-digital converter (ADC), and then output to digital-to-analog converter (DAC) after being processed by field programmable gate array

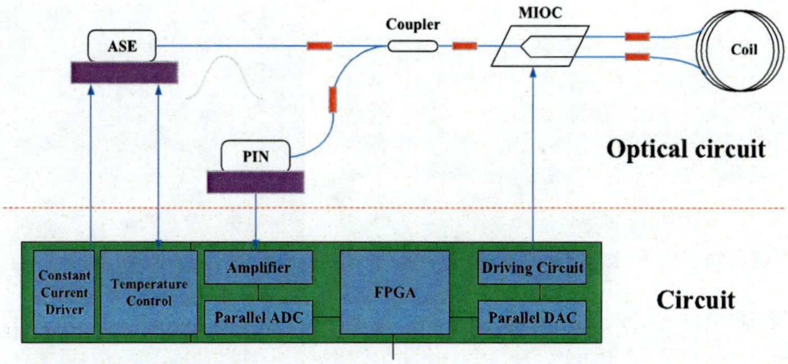


**Fig. 28.2** Schematic diagram of the single axis IFOG

(FPGA) and loaded on multi-functional integrated optical chip (MIOC) after being filtered by differential filter, which is used to realize closed-loop control. Therefore, in order to achieve the purpose of high precision and high bandwidth, the optical path and circuit are optimized in detail.

## 28.3 Configuration of Space Grade Single Axis IFOG

### *28.3.1 ASE Light Source*

In order to improve the signal-to-noise ratio of the light signal arriving at the detector, an amplified spontaneous emission (ASE) light source with a power of 5 mW is carried out, of which the scheme is shown in Fig. 28.4. In this scheme, the ASE light source including pump laser (PUMP), wavelength division multiplexer (WDM), erbium-doped fiber and fiber reflector (FR). The length and erbium ion concentration are carefully designed and optimized to meet the needs of high power and high stability. The experimental results show that all the parameters meet the application requirements of fiber optic gyroscope. The spectral type and specific test indicators are shown in Fig. 28.3 and Table 28.1, respectively.

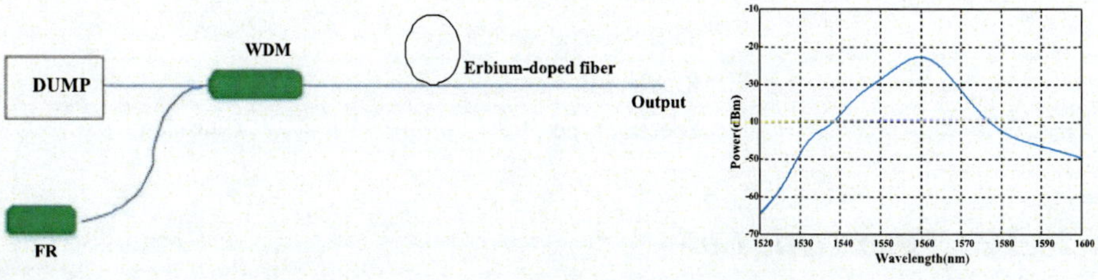


**Fig. 28.3** The scheme of ASE light source (Left: Scheme; Right: Gauss spectroscopy)

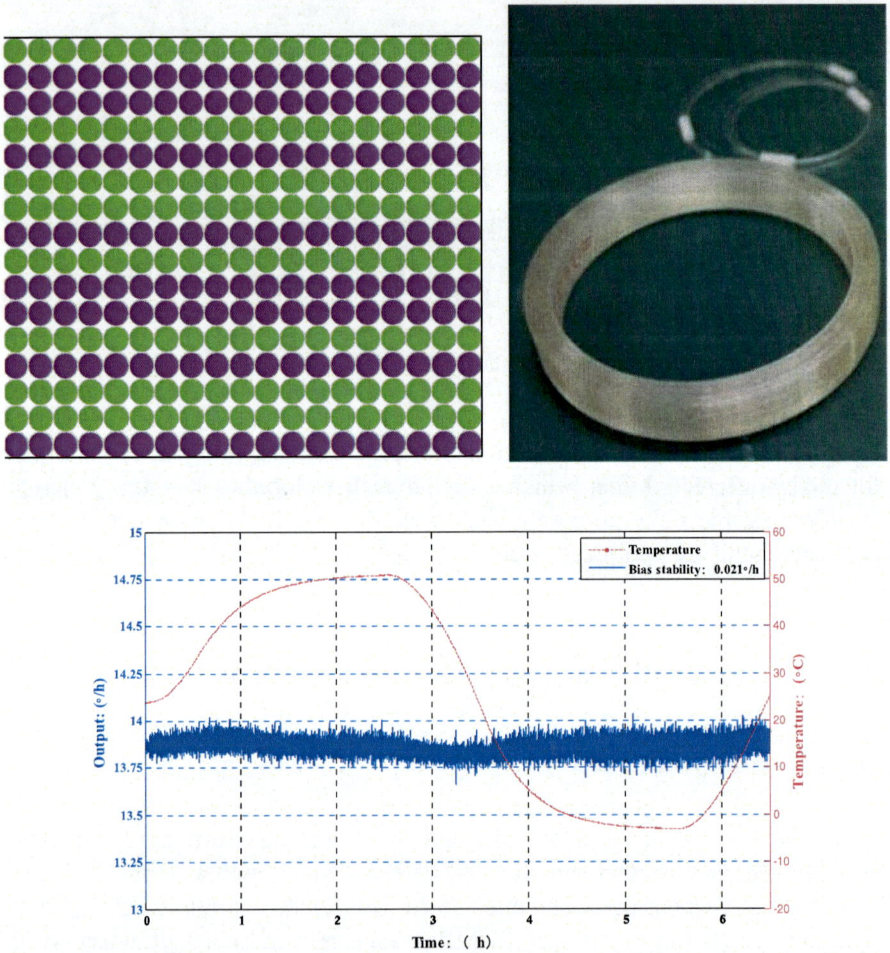


**Fig. 28.4** Schematic diagram of fiber optic coil and test curve

**Table 28.1:** Parameters of ASE light source (@100 mA)

| Parameter | results |
|---|---|
| Power, mW | ≥5 |
| Average Wavelength, nm | 1530 |
| 3 dB Spectral Width, nm | ≥10 |
| Ripple, dB | ≤0.2 |
| Spectral type | Gaussian |

## *28.3.2 Fiber Optic Coil*

In recent years, different winding methods of fiber optic coil show different characteristics for the environmental adaptability of IFOG. The fiber optic coil with good symmetry is of great significance in reducing the influence of temperature gradient for application of high precision. At present, the octupole symmetrical winding method is usually used in the optic fiber coil encircling system, which has a certain impact on the performance of high-precision IFOG due to the residual temperature error. The octupole-winding method which shown in has more obvious advantages and can further reduce the influence of temperature residual.

Figure 28.4 shows the temperature characteristic curve of fiber optic coil for octupole-winding. It is found that when the temperature range is –10 to 50 °C, the fiber optic coil with octupole-winding has obvious advantages, and the bias stability over –10 to 50 °C is around 0.021°/h.

### 28.3.3 MMS Performance

In order to enhance the performance of MMS, fundamental limitation is reduced as much as possible during our process. It can be summarized as follows:

Octupole-winding technology is used to further improve the environmental adaptability of the fiber ring with 3.8 km.

The whole machine structure thermal design is convenient to improve the product symmetry and reduce the influence of temperature gradient.

The software algorithm based on FPGA is optimized, and the bandwidth is up to 2 kHz.

Software filtering and hardware filtering are combined to reduce the noise of the product.

After optimization, single axis IFOG with different precision is designed to fulfill the requirement of as more as more missions by our team. For example, FOG-1-1 series single IFOG will be used in China's Mars exploration due to its miniaturization. FOG-2-1 series single axis IFOG has important application value in the field of commercial satellite and communication satellite, and has been widely verified in "Superview-1" satellite. In the early stage of design, considering the application requirements of remote sensing satellite, FOG-3-1 series single axis IFOG has carried out the optimization of high bandwidth and high precision, and the communication frequency can reach up to 2 kHz. The Allan variance curves of the three single axes IFOG are shown in Fig. 28.5. Aiming at the requirement of different spacecraft, different series of MMS products are developed after optimization, and the main indicators are shown in Table 28.2. All the single axis gyros can be used in the aerospace field and can be configurable for different configurations as 3S, 3S + 1 and 6S structure.

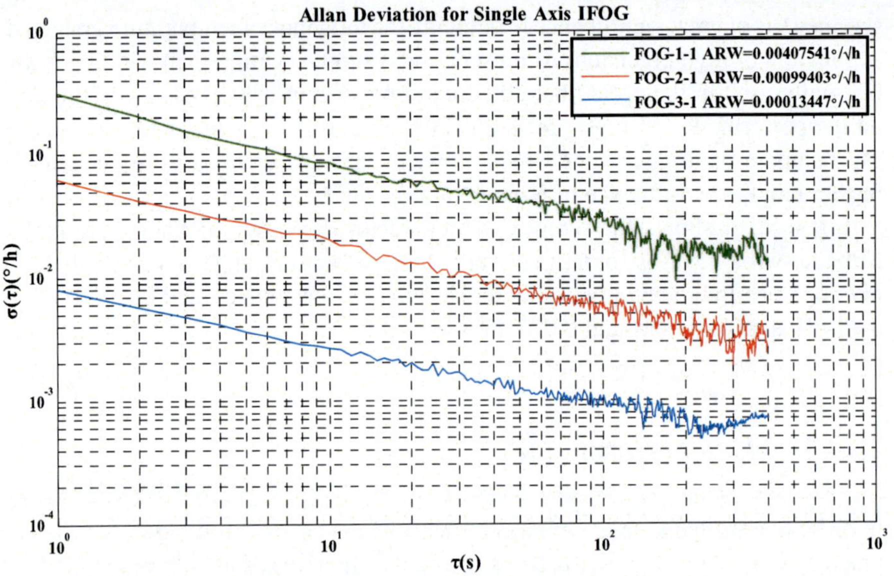


**Fig. 28.5** Single Axis fiber optic gyroscope (Allan variance)

**Table 28.2:** Parameters of single axis series

| Parameter | Single axis FOG-3-1 | Single axis FOG-2-1 | Single axis FOG-1-1 |
|---|---|---|---|
| Range of measured angular rate, °/s | ±20 | ±110 | ±200 |
| Bias drift, °/h | 0.001 | 0.01 | 0.1 |
| Scale factor repeatability, ppm | 10 | 50 | 100 |
| Bandwidth, kHz | Up to 2 | Up to 2 | Up to 2 |
| Random walk, °/√h | 0.0002 | 0.001 | 0.01 |
| Weight, kg | 1.2 | 0.42 | 0.28 |
| Dimensions, mm | Φ120 × 56 | 98 × 98 × 52 | 70 × 70 × 61.5 |

## 28.4 Conclusion

In conclusion, development of broadband sensors based on fiber optic gyroscope is carried out and further optimization will be carried out to fulfill the requirement of specific customer demands. Compared with other countries' products, our products have maintained a high-cost performance, and will play an increasingly important role in China's aerospace missions in future.

**Acknowledgements** This work was supported by a project numbered 41417060102.

## References


1. Ge, S., Yang, R., Guo, C.: Constant small tension control for fiber optic coil variable-velocity winding. IEEE Access **7**, 172012–172020 (2019)
2. Chen, S., Lin, T., Shih, Y., Hsu T.: Dynamic characteristic modeling and simulation for a fiber-optic gyroscope integration system. In: 2019 International Conference on Engineering, Science, and Industrial Applications (ICESI), pp. 1–6, Tokyo, Japan (2019)
3. Tehrani, M.M., Soltanieh, S.: Sources of bias instabilities (1/f Noise) in optical gyros. IEEE Sensors J. **22**(19), 10554–10559 (2019)
4. Wang, H., Lu, Y., Lu, T., Liu, R., Liaw, S.: Superfluorescent fiber source with ultra-low thermal coefficiency operating in the conventional band. In: 2018 Conference on Lasers and Electro-Optics Pacific Rim (CLEO-PR), pp. 1–2 (2018)
5. Yao, J., Li, K., Li, B., Wang, C., Chen, K., She, X.: Analysis of spectrum properties of integrated optical chips applied on IFOG. In: 2018 Progress in Electromagnetics Research Symposium (PIERS-Toyama), pp. 827–832 (2018)
6. Gao H., Gao W., Wang G., Sun J.: Research on optimum modulation phase of interferometric fiber optic gyroscope in the space radiation. 2018 DGON Inertial Sensors and Systems (ISS), pp. 1–11 (2018)
7. Maxime, J.R., Rinehart, S.A., Dhabal, A., et al.: The Balloon Experimental Twin Telescope for Infrared Interferometry (BETTII): towards the first flight, Publications of the Astronomical Society of the Pacific, vol. 941 (126) (2016)

# Chapter 29
# Short-Term Wind Speed Prediction Based on Phase Space Reconstruction and Elman Network-Grey Wolf Optimization

**Shengqing Li, Haotian Feng, and Xi Deng**

**Abstract** Aiming at the problem of low accuracy of short-term wind speed prediction caused by sudden changes in wind speed. A combined short-term wind speed prediction algorithm based on the phase space reconstruction method and the gray wolf algorithm to optimize the Elman network is proposed. Firstly, the algorithm uses the phase space reconstruction method to reconstruct the original sequence data, and it constructs the time series prediction model. Then, the Elman network is used to train the model and the gray wolf algorithm is used to optimize the network parameters. Finally, evaluation index effected by the prediction is proposed to evaluate the prediction effect. Through simulation and testing, this method can better improve the prediction accuracy.

## 29.1 Introduction

### *29.1.1 Background*

With the increasing depletion of fossil energy and the severe challenges posed by environmental problems, wind power as a clean and sustainable energy source, is becoming the main future source of energy. However, due to the change of wind power, its large-scale interconnection easily leads to the stability decline of grid voltage, frequency, power angle, and so on, which would bring potential safety hazards to production and daily life [1–3]. By predicting the future wind power ahead of time, it can control the predicted power ahead of time and making the generation plan. Power predication of short-term wind can effectively prevent power grid stability problems that are caused by power fluctuation of wind.

Wind power prediction research now is basically mature. There are two main predication methods: physical method and statistical method [4]. Firstly, the physical method establishes the meteorological model of the wind power plant by investigating the physical factors, such as, topography, climate background and so on. Then

S. Li · H. Feng (✉) · X. Deng
Hunan University of Technology, Zhuzhou 412007, Hunan, China

 
L. C. Jain et al. (eds.), *Smart Communications, Intelligent Algorithms and Interactive Methods*, Smart Innovation, Systems and Technologies 257,
https://doi.org/10.1007/978-981-16-5164-9_29

the meteorological information is predicted by the meteorological model. Finally, the output power of the fan is obtained by taking the meteorological information as an independent variable and the output curve of the fan as a function [5]. Methods in statistics is based on statistical theory. Firstly, historical operation data of wind power plant are collected including historical wind speed and historical power. Then the mapping relationship between them is analyzed. Finally, future power is predicted by learning this mapping relationship. There are several methods employed to analyze this mapping relationship such as time series method [6], Kalman method [7], machine learning [8, 9], support vector machine [10–17], grey prediction [18, 19].

These methods have their advantages and disadvantages, and the prediction accuracy is also different [20, 21]. This paper solves the problem of improving the accuracy of the wind power forecasting. Combined forecasting algorithm based on Elman network-grey wolf optimization is proposed. And its effectiveness is verified by the modeling and analysis of MATLAB 2015.

### *29.1.2 Phase Space Reconstruction*

Nowadays, wind power plant can only collect meteorological information for example wind speed, wind direction, air pressure and temperature. However, for non-linear systems, the time series of one or several components can not meet the input requirements of the system. Besides, the degree of information correlation in time dimension is not considered, which often results in the loss of input information. On account of this, the input of the system can be reconstructed to increase the input information saturation. Embedding dimension h and time delay p are two important parameters for phase space reconstruction. Two important indicators of phase space reconstruction are the standard embedding dimension h and the time delay $p$.

Based on the appropriately determined values of h and p, if the time series of wind speed is $\{x(t)|t = 1, 2, \ldots, N\}$, the reconstructed wind speed sequence would be

$$Y(t) = [x(t), x(t+p), x(t+2p), \ldots, x(t+(h-1)p)],\ t = 1, 2, 3, \ldots, N-(m-1)p \quad (29.1)$$

Among them, G-P (Grassberger-Procaccia) algorithm is used to determine the value of h. And p is usually determined by autocorrelation function method, average mutual information method and reconstruction expansion method. The pseudo-phase space is composed of $Y$ and is recorded as $R^m$. The evolution rule of $R^m$ midpoint $Y(t) \to Y(t+p)$ is the same as that of the original system, and many features in that system can be reflected in the reconstruction space.

Formula (29.1) is taken as input and $Y(t+p)$ as output to establish prediction model. The mapping relationship between input and output is determined to realize wind speed prediction.

## 29.2 Elman Network

A typical recurrent neural network calls elman neural network (RNN), as shown in Fig. 29.1. Its internal feedback and feedforward structure' equips it with stronger learning ability than ordinary neural network. It is very suitable for building a non-linear model of wind speed prediction.

In the learning algorithm of Elman network $S(k)$ and $y(k)$ are as follows:

$$S(k) = f(W_1X + W_3S(K-1) + B_1) \tag{29.2}$$

$$y(k) = \sum W_2S(k) + B_2 \tag{29.3}$$

In formula (29.2)–(29.3), $k$ represents time; $f$ denotes the transfer function of the hidden layer and generally takes $S$-type function; $S(k-1)$ is the feedback term of the hidden layer. The offset values of the input layer $B_1$ and the hidden layer $B_2$. The weights of the input layer $W_1$, the hidden layer $W_2$, and the output layer $W_3$. The input values $X$ and the output values $y$.

The global approximation error is:

$$E = \frac{1}{2}\sum_{K=1}^{T}[y(k) - y^0(k)]^2 \tag{29.4}$$

In formula (29.4): $y^0(k)$ represents the target output at k time.

The modification of weights $W_1$ and $W_2$ is similar to that of feedforward neural networks. The main difference lies in the weight $W_3$. In reference [10], a chain rule is proposed to obtain the weights, e.g. (29.5):

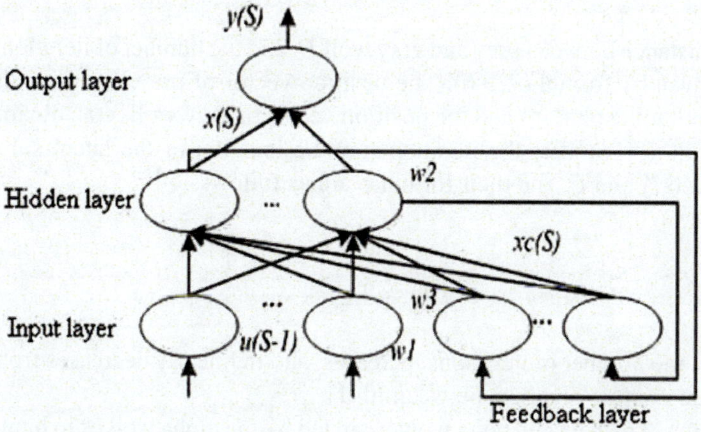


**Fig. 29.1** Elman neural networks structure

$$\Delta W_3 = \sum_{k=1}^{T} [(y^0(k) - y(k))W_2]f'(S(k))S(k-1) \tag{29.5}$$

## 29.3 Grey Wolf Algorithm

### *29.3.1 Rationale of Grey Wolf Algorithm*

The main wolf is marked as alpha ($\alpha$), and the remaining grey wolves are recorded as beta, delta and omega ($\beta$, $\delta$ and $\omega$) in the algorithm. The alpha wolves are always in the best fitness position and the closest to the optimal value, while the beta and delta as the candidates for alpha wolves have their second best fitness, and the rest are omega wolves. In the whole algorithm process, alpha wolves are responsible for leading wolves to search, track and approach prey, beta and delta wolves attacking prey, and omega wolves hunting prey.

### *29.3.2 Description of Grey Wolf Algorithm*

The algorithm can divide the whole hunting process into three steps: encircling, chasing and attacking, and ultimately capture the prey (perform the optimization).

Encircling: The distance between prey and wolf during this period.

$$\left\{ \begin{array}{l} D = C * X_p(t) - X(t) \\ X(t+1) = X_p(t) - A * D \end{array} \right\} \tag{29.6}$$

The distance between prey and gray wolf $D$, and the number of iterations $t$, respectively, given by formula (29.6). The position vector of prey $X_p(t)$ (i.e. the location of the optimum solution) and the position vector of grey wolf $X(t)$, obviously, given by formula (29.6) after the iteration (i.e. the location of the latent solution). The coefficients $A$ and $C$, and their formulas are as follows:

$$\left\{ \begin{array}{l} A = 2m * r_1 - m \\ C = 2r_2 \end{array} \right\} \tag{29.7}$$

When the number of iterations increases, and m linearly decreases from 2 to 0, $r_1$ and $r_2$ are random numbers between [0, 1].

Chasing: The beta and delta wolves are led by the alpha wolves to hunt their prey. The position of the prey (optimal solution) is re-determined. The renewal equation is as follows:

$$\begin{cases} D_\alpha = |C_1 X_\alpha(t) - X(t)| \\ D_\beta = |C_2 X_\beta(t) - X(t)| \\ D_\delta = |C_3 X_\lambda(t) - X(t)| \end{cases} \tag{29.8}$$

$$\begin{cases} X_1 = X_\alpha(t) - A_1 D_\alpha \\ X_2 = X_\beta(t) - A_1 D_\beta \\ X_3 = X_\lambda(t) - A_1 D_\lambda \end{cases} \tag{29.9}$$

$$X_p(t+1) = \frac{X_1 + X_2 + X_3}{3} \tag{29.10}$$

Among them, $D_\alpha$, $D_\beta$, $D_\lambda$ represent the distance between the wolves $\alpha$, $\beta$, $\lambda$ and the wolves $\omega$ (other individuals).

Attacking: It is realized by decreasing the value of $a$ in Eq. (29.10). When $A$ are in $[-1, 1]$, it indicates that wolves would converge towards the prey, while when $1 < |A| \leq 2$, wolves would diverge from the prey, which would deprive the grey wolf algorithm of its optimal position, thus falling into a local optimal process.

### *29.3.3 Grey Wolf Optimization Algorithm*

See Fig. 29.2.

## 29.4 Wind Speed Combination Predication and Analysis

### *29.4.1 Prediction Process*

The traditional wind speed prediction model takes the single-dimensional wind speed data as the input, which causes a great loss of implicit information. However, the wind speed in the adjacent time period would affect the trend of its future wind speed change. By means of autocorrelation method and G-P method, this paper takes the maximum value of future wind speed impact factor as the objective function to investigate and solve the influence of historical wind speed on future wind speed in different phases. The reconstructed wind speed input can not only effectively release the hidden information in the wind speed sequence, but also maximize the learning ability of Elman network with appropriate dimension input.

In order to effectively improve the learning ability and learning rate of the Elman network, this paper uses the gray wolf optimization algorithm to optimize the initial weights and thresholds of the network. In the model training stage, through batch training process, the weights and thresholds of $n$ groups are obtained, and they are all optimal weights and thresholds; then the weights and thresholds of $n$ groups

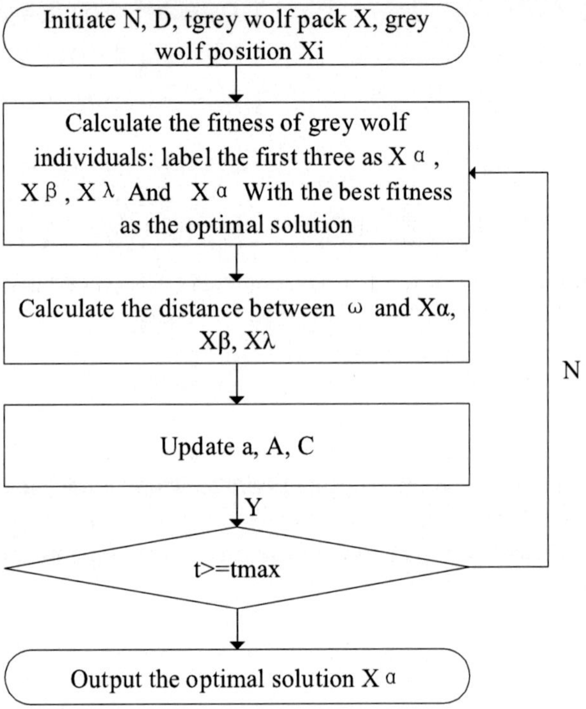


**Fig. 29.2** Grey wolf optimization algorithm

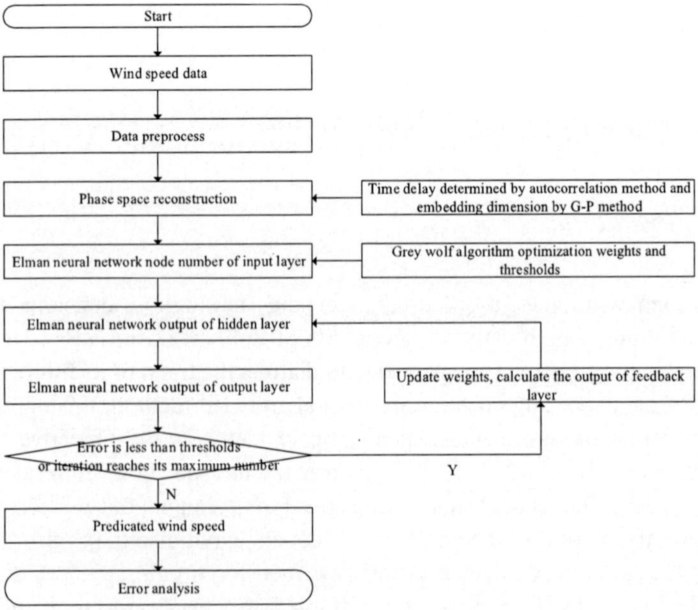


**Fig. 29.3** Flow chart of combined model for wind speed prediction

are cross-validated, and the optimal group of weights and thresholds is selected for the test stage. Through cross-validation, the generalization ability of weights and thresholds for different batches of inputs can be well tested, and a group of weights and thresholds with the strongest generalization ability can be screened as well.

By optimizing the input, weight and threshold, a combined model of short-term wind speed prediction based on Elman network-Grey wolf optimization is proposed in Fig. 29.2 (Fig. 29.3).

### 29.4.2 Error Evaluation Function

The estimation of prediction error is evaluated by three indexes: *MSE*, *MAE* and maximum absolute error.

$$\begin{cases} MSE = \sqrt{\sum (x_{实际} - x_{预测})^2 / n} \\ MAE = \sum \left|(x_{实际} - x_{预测})\right| / n \end{cases} \tag{29.11}$$

In the formula, $n$ is the data value and $x$ is the wind speed vector.

### 29.4.3 Simulation Experiment and Result Analysis

The original data is the actual wind speed of a wind farm from October 1, 2016 to December 31, 2017. With 15 min as a sampling point, nearly 30,000 wind speed data were collected. After preprocessing, 16,213 pieces of available data are obtained, as shown in Fig. 29.4. From the wind speed curve, they cover all the wind speed sections, which can be regarded as a set of ideal simulation data to reflect the actual change.

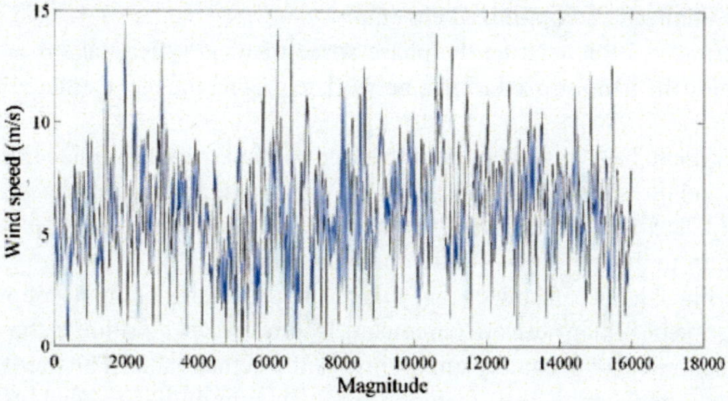


**Fig. 29.4** Wind speed of wind field

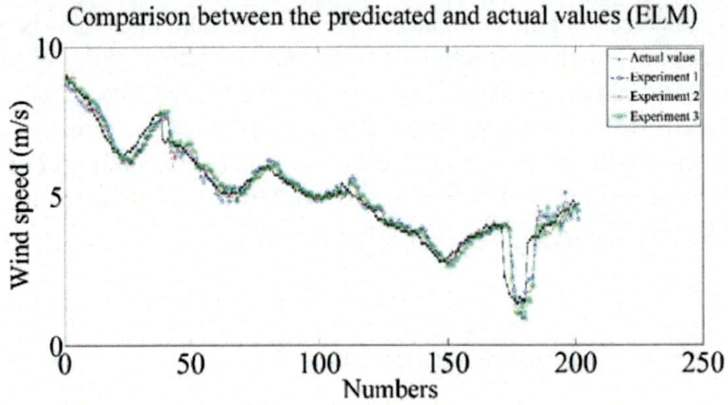


**Fig. 29.5** Comparison of predicted wind speeds in different experiments

Input and Output: The original data is cleaned and reconstructed. As the final input of Elman network, $p$ is determined to be 15 by autocorrelation method and $h$ is determined to be 6 by G-P algorithm. Therefore, the output and input data of each wind speed are column vectors of 1 * 6, i.e. input and output is as formula (29.12):

$$\begin{cases} X = [x(t), x(t+15), x(t+30), x(t+45), x(t+60), x(t+75)] \\ Y = x(t+90) \end{cases} \tag{29.12}$$

Training phase: Take the first 15,000 cleaned wind speeds as training data, and use the Grey wolf algorithm to optimize the network.

Test phase: Take 1000 cleaned wind speed as test data. Through three groups of different experiments, the prediction effect of the combined algorithm is compared and analyzed.

Experiment 1 Without pretreatment of wind speed, that is, without phase space reconstruction, the input of Elman network becomes a single wind speed of 1 * 1. The output remains unchanged, and the weights and thresholds of Elman network are not initialized and optimized either.

Experiment 2 Reconstruct the phase space of wind speed, take $p = 15$, $h = 6$, and employ the unoptimized Elman network to predict the reconstructed wind speed data.

Experiment 3 Reconstruct the phase space of wind speed, optimize Elman's initial weights and thresholds by using grey wolf algorithm, and predict the wind speed.

Figure 29.5 is a comparison of the predicted and actual values of wind speed obtained from three groups of different experiments.

From Fig. 29.5, it can be seen that with the preprocessing of input, coupled with the initialization of Elman network parameters by Grey wolf algorithm in the early stage, the prediction curve gradually approximates the actual value. The prediction curve is smooth, and when there is a significant fluctuation in wind speed, its prediction can also maintain a high stability.

**Table 29.1** Comparison of three types of wind power forecasting errors

| Experiment | MSE | MAE |
|---|---|---|
| One | 0.4064 | 0.2506 |
| Two | 0.3042 | 0.1823 |
| Three | 0.2239 | 0.1279 |

According to Table 29.1, the MSE and MAE in experiment 3 are 0.2239 and 0.1279, which are the minimum error values of previous experiments. It shows that the wind speed prediction model based on Elman network-grey wolf optimization can more accurately predict the wind speed process, and the effect is better than the first two experiments.

On the basis of the above simulation results, by reconstructing the appropriate input and using the optimized Elman network, which can enhance the accuracy of wind speed prediction. The combined model of short-term wind speed prediction based on Elman network-grey wolf optimization can greatly improve the accuracy of prediction.

## 29.5 Conclusion

Aimed at the problem of poor prediction accuracy caused by the randomness of wind speed, the appropriate input is reconstructed by phase space, and combined with the optimized Elman network to predict wind speed, which greatly improves the prediction accuracy. The conclusions are as follows:

(1) Data reconstruction can effectively display and magnify the hidden information of the data, and increase the information input of the prediction model.
(2) By optimizing the initial weights and thresholds of Elman network through grey wolf algorithm, the prediction speed and convergence rate of the network can be effectively enhanced and the prediction accuracy of the network can be reached as well.
(3) The combined model of short-term wind speed predication based Elman network-grey wolf optimization has better prediction accuracy than the single model, and MSE, MAE of prediction results are reduced as well.

## References

1. Cheng, S.Y., Song, S.F., Li, L.X.: Survey on smart grid technology. Power Syst. Technol. **8**, 40–43 (2009)
2. Lange, B., Rohrig, K., Ernst, B.: Wind power prediction in Germany-recent advances and future challenges. In: European Wind Energy Conferences, D-34119 Kassel, Germany (2016)

3. Lange, M., Focken, U., Heinemann, D.: Previento-regional wind power prediction with risk control. In: World Wind Energy Conference, Berlin Germany (2012)
4. Qian, Z., Pei, Y., Cao, L.X.: Review of wind power forecasting method. High Voltage Eng. **4**, 1047–1060 (2016)
5. Zhu, X.R., Liu, Y.P.: Wind power forecasting using time series model based on robust estimation. Proc. CSU-EpSA **3**, 103–110 (2012)
6. Zhou, S.L., Miao, M.Q., Su, J.W.: Prediction of wind power based on principal component analysis and artificial neural network. Power Syst. Technol. **4**, 19–23 (2014)
7. Pan, D.F., Liu, H., Li, Y.F.: Optimization algorithm of short-term multi-step wind speed forecast. Proc. CSEE **26**, 87–91 (2008)
8. Cui, Q., Jie, S., Zhang, X.: The application of improved BP neural network for power load forecasting in the island microgrid system. In: International Conference on Electrical & Control Engineering, Yichang China (2011)
9. Zhang, L.Y., Meng, L.J., Wang, Z.Z.: Photovoltaic power station output power prediction based on double Bp neural network. Electr. Meas. Instrum. **11**, 31–35 (2015)
10. Li, R., Ke, Y.Q., Zhang, X.: G: Wind power forecasting based on time series and SVM. Electr. Power **1**, 64–68 (2012)
11. Ding, Z.Y., Yang, P., Yang, X.: Wind power prediction method based on sequential time clustering support vector machine. Autom. Electr. Power Syst. **14**, 131–135 (2012)
12. Wang, K., Guan, S.Q., Wang, L.X.: A combined forecasting model for wind power prediction based on fuzzy information granulation and least squares support vector machine. Power Syst. Prot. Control **2**, 26–32 (2015)
13. Li, X., Wang, X., Zhen, Y.H.: Short-term wind load forecasting based on improved LSSVM and error forecasting correction. Power Syst. Prot. Control **11**, 64–69 (2015)
14. Chen, D.J., Gong, Q.W., Jin, C.Y.: Short-term wind power prediction based on support vector regression machine optimized by adaptive disturbance quantum-behaved particle swarm optimization. Power Syst. Technol. **4**, 974–980 (2013)
15. Yang, Z.L., Liu, Y.: The prediction of short-term wind power by applying particle swarm optimization. Power Grid Technol. **5**, 159–164 (2011)
16. Wang, Y.X., Chen, G.C.: Short-term wind power prediction based on IAFSA optimization SVM. Electr. Meas. Instrum. **3**, 80–84 (2016)
17. Ye, L., Liu, P.: Combined model based on EMD-SVM for short-term wind power prediction. Proc. CSEE **31**, 102–108 (2011)
18. Wang, L.J., Dong, L., Liao, X.: Z: Short-term power prediction of a wind farm based on wavelet analysis. Proc. CSEE **28**, 30–33 (2009)
19. Meng, X.X., Tian, C.W., Dong, L.: A study of grey theory used in prediction of medium and long-term wind power generation. Power Syst. Prot. Control **21**, 81–85 (2011)
20. Meng, Y.F., Hu, S.J., Deng, Y.: Analysis and evaluation method of wind power prediction-error. Power Constr. **7**, 6–9 (2013)
21. Li, K., Gao, Y.: A combination method for wind power prediction based on cooperative game theory. Electricity (Z1), **2**, 36–40 (2014)

# Chapter 30
# Modern-Driven Deep Learning for Belief Propagation LDPC Decoding

**Yiduo Tang, Lin Zhou, Shuying Zhang, and Yan Chen**

**Abstract** In this paper, for the long code decoding problem, we analyze the performance of belief propagation (BP) decoder in neural network. The decoding of long codes has always been a concern of LDPC decoding. In recent years, the application of neural networks in the communication field has gradually become widespread. As a result, we are considering and combining the two. The decoding method proposed in this paper uses model-driven deep learning. The network we propose is a neural standardized BP LDPC decoding network. Model-driven deep learning absorbs the advantages of both model-driven and data-driven, which combines them adaptively. The network structure proposed in this paper takes advantage of model-driven to expand the iterative process of decoding between check nodes and variable nodes into the neural network. We can increase the number of iterations by increasing the CN layer and VN layer of the hidden layer. Furthermore, by changing the SNR to detect its relationship with system robustness, and, finally, determine the appropriate SNR range.

## 30.1 Introduction

LDPC codes which are a well-known linear error correcting codes were first proposed by Gallager in 1962 [1]. However, due to the limitation of computing power and hardware development level at that time, the code did not attract attention at that time. Since the 1996 literature [2] rediscovered LDPC codes, this code has attracted the attention of scholars together with Turbo codes due to its excellent error correction performance approaching the Shannon limit and its high decoding throughput and has become a new research hotspot.

Y. Tang · L. Zhou · S. Zhang · Y. Chen (✉)
Xiamen Key Laboratory of Mobile Multimedia Communications, National Huaqiao University, Xiamen 361021, Fujian, China
e-mail: goldency@hqu.edu.cn

L. Zhou
State Key Laboratory of Integrated Services Networks, Xidian University, Xi'an 710071, Shaanxi, China

 
L. C. Jain et al. (eds.), *Smart Communications, Intelligent Algorithms and Interactive Methods*, Smart Innovation, Systems and Technologies 257,
https://doi.org/10.1007/978-981-16-5164-9_30

So far, belief propagation (BP) decoding algorithm is one of the main LDPC decoding algorithms. The computational complexity of the BP decoding algorithm is related to the code length of the LDPC code. It can be applied in a variety of environments and can achieve good performance.

Researchers have been conducting in-depth research and improvement on LDPC encoding and decoding algorithms. Deep learning has powerful learning and extracting feature attributes, which is suitable for dealing with high complexity problems. In recent years, researchers have gradually begun to use deep learning to solve decoding problem in communication. It has been illustrated that assigning appropriate weights to the transmission information in Tanner graph can reduce iterations of the BP decoder with identical decoding performance [3]. In [4], author proposed a decoding method which used neural min-sum to reduce the complexity of decoding in hardware.

In order to design a decoder for LDPC long code with lower complexity, we propose a new neural BP LDPC decoding network. Iterative decoding is performed by unfolding the BP of long code decoding into a feedforward propagation network.

## 30.2 Related Work

### *30.2.1 BP Decoding*

LDPC codes are a kind of $(N, K)$ linear block codes with good forward error correction performance. Among them, the parity check matrix $\mathbf{H}$ is sparse, and the number of non-zero elements in the matrix is much smaller than the number of zero elements. For $(N, k)$ LDPC codes, the code length is represented by $N$, which value is equal to the total number of columns in the check matrix. $M$ represents the length of the check sequence, which value also shows the total number of rows of the check matrix. $K$ represents the information sequence length.

In addition to using the check matrix to represent an LDPC code, the graph model is another way to represent LDPC codes. The Tanner graph is a common method in the graph model.

The Tanner graph is only composed of check nodes, variable nodes, and the connection between the two, and the connections are all undirected [5], in which each check node is summed one row of the check matrix corresponds to each other, and each variable node corresponds to a column of the check matrix. Therefore, the Tanner graph includes $M$ check nodes and $N$ variable nodes in total, and when the element $h$ in the $i$th row and the $j$ column of the check matrix is $h = 1$, there is a line between the $i$th check node and the $j$th variable node in the Tanner graph.

In the BP decoding process, the messages in the channel are transmitted iteratively using the variable nodes and check nodes of the Tanner graph. We use $r_i$ define channel output. Let $L(r_{ji})$ denotes the messages from CN to VN, $L(q_{ij})$ denotes messages from VN to CN, and $L(P_i)$ denotes the log-likelihood ratio (LLR) of the receiving channel corresponding to bit $i$. The information delivery process is as follows:

Step 1 The channel output signal $r_i$ send to the decoder and then decoder calculates the corresponding LLR, the $n$th coded bit $w_i$ as follow:

$$L(q_{ij}) = L(P_i) = \log \frac{\Pr(w_i = 0|r_i)}{\Pr(w_i = 1|r_i)} = \frac{2r_i}{\sigma^2} \quad (30.1)$$

where $\sigma^2$ is the channel noise power following the Gaussian distribution.

Step 2 Message updated in VN:

$$L(r_{ji}) = 2\tanh^{-1}\left(\prod_{i' \in R_j \backslash i} \tanh\left(\frac{1}{2}L(q_{i'j})\right)\right) \quad (30.2)$$

Step 3 Message updated in CN:

$$L^{(l)}(q_{ij}) = L(P_i) + \sum_{j' \in C_i \backslash j} L^{(l)}(r_{j'i}) \quad (30.3)$$

Step 4 Calculate the sum of LLR:

$$L(Q_i) = L(P_i) + \sum_{j \in C_i} L(r_{ji}) \quad (30.4)$$

If the convergence condition is met or the number of iterations reaches the threshold, the operation ends. If not, repeat from step 2. Then, calculate hard decision messages for all variable nodes. If $L(Q_i) > 0$, then the estimated bit value $o_i = 0$, otherwise $o_i = 1$.

### 30.2.2 Model-Driven Deep Learning

Deep learning is a standard data-driven approach, which takes deep network as a black box and relies on a large amount of data to solve real problems. The model-driven method starts from the priory of the target mechanism to form a learning cost function and then minimizes the cost function to solve the problem. In the field of communication, model-driven deep learning can build network architecture based on existing communication models. At present, this method has been successfully used in communication, including the improvement of channel decoding algorithm [6, 7].

## 30.3 Neural BP Structure

In this chapter, we will introduce how to realize BP decoding in neural network and the detailed design information of each layer of the network.

First, Fig. 30.1 gives the architecture of the proposed decoding network. The code word $x$ of length $k$ is encoded into the binary code word $u$ of length $N$ and then modulated by BPSK. The BPSK symbol will be passed through a channel with Gaussian noise. At the receiver, use the receiver signal $y = s + n$ to calculate the LLR value and then feed it into the deep feedforward neural network.

Then, we give a detailed signal flowchart on the receiver side of the decoding network structure, as shown in Fig. 30.2. The iterative decoding algorithm is expanded into a forward feedback network. Each hidden layer corresponds to a BP iteration process.

The hidden layer contains two types of neurons: one is the CN neuron which calculates the extrinsic information in the CN layer:

$$L^{(l)}(r_{ji}) = 2\tanh^{-1}\left(\prod_{i'\in R_j\backslash i}\tanh\left(\frac{1}{2}L^{(l-1)}(q_{i'j})\right)\right) \tag{30.5}$$

Another one is VN neuron. It calculates the extrinsic information in the VN layer:

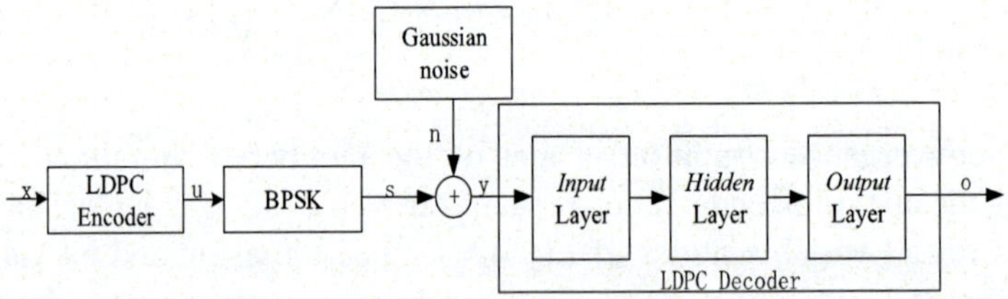


**Fig. 30.1** Process diagram of the decoding algorithm for neural BP network

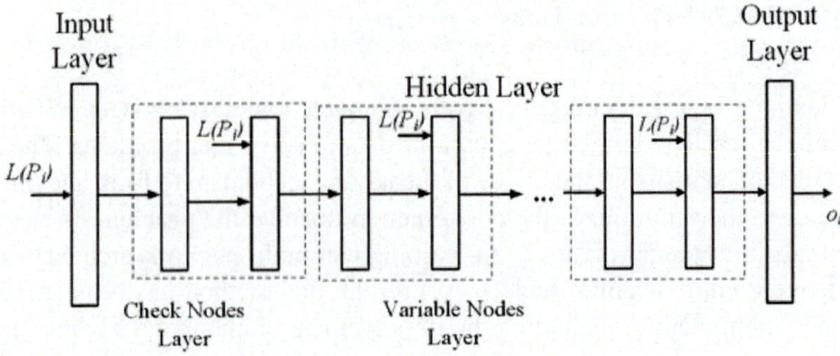


**Fig. 30.2** Signal flowchart of model-driven deep learning BP decoding network

$$L^{(l)}(q_{ij}) = L(P_i) + \sum_{j' \in C_i \backslash j} L^{(l)}\left(r_{j'i}\right) \tag{30.6}$$

In the output layers, there is:

$$o_i = \sigma\left(L(P_i) + \sum_{j \in C_i} L^{(l_{\max})}\left(r_{ji}\right)\right) \tag{30.7}$$

We use $\sigma = \left(1 + e^{-x}\right)^{-1}$ as activate function in output neurons.

It is easily to increase the iterations by increasing the CN layer and VN layer of the hidden layer. Besides, we choose cross entropy as loss function. It is defined as:

$$L(P, o) = -\frac{1}{N}\sum_{i=1}^{N} P_i \log(o_i) + (1 - P_i)\log(1 - o_i) \tag{30.8}$$

## 30.4 Simulation Result and Analysis

In this section, we analyze the performance of neuron BP LDPC decoding algorithm through numerical simulation to verify the effectiveness of the algorithm. To facilitate training, we use the Tensorflow framework [8]. The training data uses the code with length 567 and rate 3/4. The parity check matrix **H** was selected from [9].

Network training uses mini-batch gradient descent. There are 120 examples in each mini-batch. In addition, we applied the Adam optimization method for searching the optimal network parameters with learning rate equals 0.03.

In order to test the robustness of the decoder, we choose to use varying SNR to generate training data. SNR equal to 0 dB means that the channel condition for data generation is poor, and many errors are generated in the data set, which will make the network unable to correctly learn the structure of the decoder, resulting in under-fitting of the network. SNR equal to 5 dB means that the channel conditions for data generation are very good, and there will be very little error data feedback input into the network, which will make the network generalization ability poor and cannot eliminate noise correctly. As shown in Fig. 30.3, compared with using 0 dB or 5 dB SNR alone to generate training data, the performance of the decoder will be better with data trained in multiple channels conditions of 0–5 dB. Figure 30.4 shows the relationship between BER performance and the number of network layers. It can be seen that in a certain network layer, the BER performance improves with the increase of the number of iterations. But the network has more layers, the complexity is higher. Therefore, after comprehensive consideration, a 44-layer network was finally selected for 20 iterations.

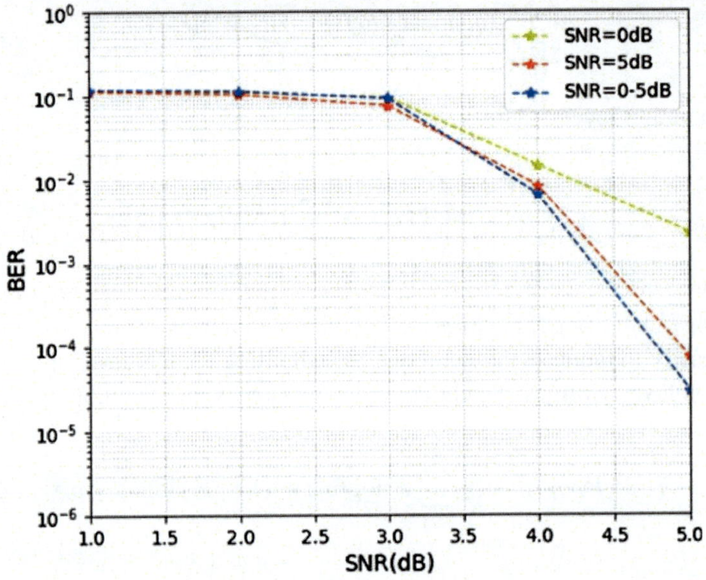


**Fig. 30.3** BER performance under different SNR conditions

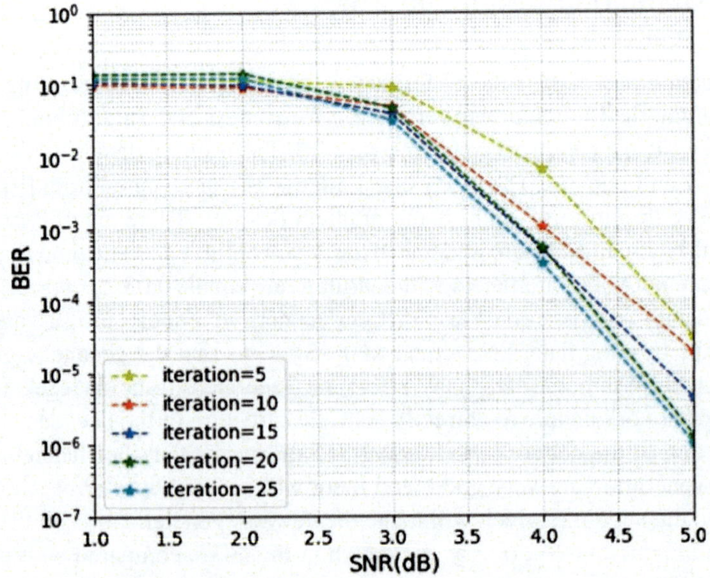


**Fig. 30.4** BER performance under different number of network layers

## 30.5 Conclusion

This article introduces the iterative unfolding process of BP decoding algorithm and then proposes an improvement scheme based on forward feedback. The improved scheme expands the iterative process of decoding into the neural network and uses the self-learning characteristics of the neural network to perform more reliable decoding. We believe that the network structure of the neural network can be used to increase the number of iterations of decoding to improve BER.

## References

1. Gallager, R.: Low-density parity-check codes. IRE Trans. Inf. Theory **8**(1), 21–28 (1962)
2. Mackay, D.J.C., Neal, R.M.: Near Shannon limit performance of low density parity check codes. Electron. Lett. **32**(18), 1645–1646 (1996)
3. Nachmani, E., Be'ery, Y., Burshtein, D.: Learning to decode linear codes using deep learning. In: 2016 54thAnnual Allerton Conference on Communication, Control, and Computing (Allerton), pp. 341–346. IEEE, Monticello, IL, USA (2016)
4. Wang, Q., Wang, S., Fang, H.: A model-driven deep learning method for normalized min-sum LDPC decoding. In: 2020 IEEE International Conference on Communications Workshops (ICC Workshops), pp.1–6. IEEE, Dublin, Ireland, Ireland (2020)
5. Tanner, R.M.: A recursive approach to low complexity codes. IEEE Trans. Inf. Theory **27**(5), 533–547 (1981)
6. He, H., Jin, S., Wen, C.-K., Gao, F., Li, G.Y., Xu, Z.: Model-driven deep learning for physical layer communications. IEEE Wirel. Commun. **26**(5), 77–83 (2019)
7. Zhang, J., He, H.: Yang X: Model-driven deep learning based Turbo-MIMO receiver. In: 2020 IEEE 21st International Workshop on Signal Processing Advances in Wireless Communications (SPAWC), pp. 1–5. IEEE, Atlanta, GA, USA (2020)
8. Abadi, M., Agarwal, A., Barham, P., Brevdo, E., Chen, Z., Cito, C., Corrado, G.S., Davis, A.: Tensorflow: Large-scale machine learning on heterogeneous systems. arXiv preprint arXiv:1603.04467 (2016)
9. Channel Codes. www.uni-kl.de/channel-codes. Last accessed 2020/11/27

# Chapter 31
# Blind Recognition of Channel Coding Based on Deep Learning

**Shuying Zhang, Lin Zhou, Zizhuo Zhao, and Yiduo Tang**

**Abstract** Blind recognition of channel coding means that the receiver can identify which channel coding type is used by the transmitter without any prior knowledge of channel coding. Traditional blind recognition algorithms for channel coding have high computational complexity and can only recognize specific coding type. In this paper, an algorithm to blindly identify channel coding types via deep learning is proposed, which can automatically extract the signal features and avoid complicated calculation. In particular, quasi-cyclic low-density parity-check (QC-LDPC) code and spatially coupled low-density parity-check (SC-LDPC) code are used as candidate codewords, and these two coding types are identified from noisy signals. Two kinds of recognizers based on long short-term memory (LSTM) network and convolutional neural network (CNN) are proposed, and the channel coding types are recognized through the classification process of neural network. Experimental results show that the proposed blind recognition method is able to realize high recognition accuracy even in poor channel conditions.

## 31.1 Introduction

Channel coding is to resist noise interference by inserting check bits in the information sequence through certain rules. LDPC code has attracted wide attention because of its performance approaching Shannon limit. In a cooperative communication system [1], the receiver can know the channel coding type and parameters adopted by the sender through the control channel. The receiver can decode correctly according to the information. In the environment of non-cooperative communication [2], the receiver must identify which channel coding is used by the sender in order

S. Zhang · L. Zhou · Z. Zhao (✉) · Y. Tang
Xiamen Key Laboratory of Mobile Multimedia Communications, National Huaqiao University, Xiamen 361021, Fujian, China
e-mail: zhaozizhuo@hqu.edu.cn

L. Zhou · Z. Zhao
State Key Laboratory of Integrated Services Networks, Xidian University, Xi'an 710071, Shaanxi, China

 
L. C. Jain et al. (eds.), *Smart Communications, Intelligent Algorithms and Interactive Methods*, Smart Innovation, Systems and Technologies 257,
https://doi.org/10.1007/978-981-16-5164-9_31

to decode the received signal correctly, and this identification is usually based on the unknown information of the sender.

In this paper, we propose two recognizers based on LSTM and CNN, which can blindly identify the channel coding type through the classification process of deep learning. In our method, the recognizer first goes through a training stage, in which, the two recognizers are trained using data with known labels. Since the training dataset is a small part of all possible codeword datasets, it may cause the recognizer to be unable to correctly recognize the coding sequence that are not exist in the training dataset. However, through a large number of experiments, we prove that using the proposed deep neural network structures, such as LSTM and CNN, we can train a recognizer with good performance without any knowledge of channel coding.

## 31.2 Problem Formulation

It is assumed that the information sequence before channel coding is $b = [b_1, b_2, \ldots, b_K]$, the sequence after channel coding is $c = [c_1, c_2, \ldots, c_N]$, usually $N > K$, and the sequence after modulation is $s = [s_1, s_2, \ldots, s_N]$. The sequence received by the receiver is $r = [r_1, r_2, \ldots, r_N]$. Denote the candidate set of codes by $\Theta = \{C_1, C_2, \ldots, C_n\}$, where $n$ is the total number of channel coding types adopted by the sender. The receiver needs to identify which encoding type in the set $\Theta$ is used by the sender according to the signal $r$ and the set $\Theta$ [3]. Another problem is that the receiver needs to judge whether $C_x$ is used by the sender according to the received sequence $r$ and the known coding scheme $C_x$, so as to decide whether $r$ can be decoded correctly by $C_x$. In this paper, we focus on solving the former.

## 31.3 Deep Learning Method

In this section, we introduce the proposed recognizer based on LSTM and CNN, and describe their training process.

### *31.3.1 LSTM Recognizer*

Recurrent neural network (RNN) is good at dealing with problems related to time series [4]. There is a certain correlation between different bits of the code word after channel coding, so RNN can be considered to solve this problem. LSTM is a variant of RNN, which overcomes the problem that the original RNN is difficult to deal with long-term dependence. Therefore, we expect it to perform well in dealing with coding sequences with long dependencies.

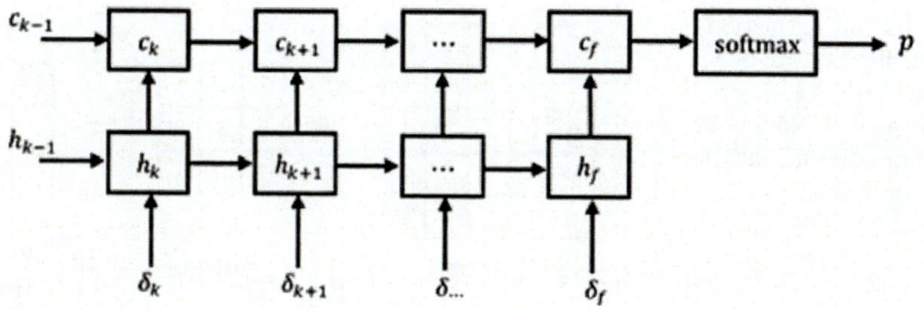


**Fig. 31.1** Structure of LSTM recognizer

The LSTM structure is shown in Fig. 31.1. There are two 'state vector' in the hidden layer of LSTM, state **h** and state **c**. State **h** is very sensitive to short-term input, and state **c** can preserve long-term state. Thanks to this structure, LSTM can deal with long-time series problems well. In this paper, we first divide the observed sequence $r$ into $f$ segments, then the length of each section is $d = N/f$, where $N$ is the length of the sequence $r$. Note that it is necessary to choose an appropriate $f$ to ensure that $d$ is an integer. Then, we input the whole observed sequence into LSTM network according to $f$ time steps.

At the kth time, LSTM received the kth sequence $\delta_k$ as input, the instant state of hidden layer $\mathbf{h}_k = g(\mathbf{h}_{k-1}, \delta_k)$ contains the current input and all the historical information of the past moment. The unit state $\mathbf{c}_k = \sigma(\mathbf{c}_{k-1}, \mathbf{h}_k)$, which determines how much of the previous unit state $\mathbf{c}_{k-1}$ and the current hidden layer state $\mathbf{h}_k$ remain to the current time $\mathbf{c}_k$. $g(\cdot)$ and $\sigma(\cdot)$ are nonlinear functions, which depend on the structure of LSTM. At last, the state $c_f$ is connected to the softmax activation function to identify the channel coding type and outputs the estimated probability **p**.

## 31.3.2 CNN Recognizer

Compared with the traditional deep neural network, CNN because of its advantages of local connection and weight sharing, which greatly improves the speed of model training. CNN is usually use to process topological data with grid structure [5]. Convolution layer and pooling layer are typical structures of CNN [6], as shown in Fig. 31.2. Convolution layer is the most important part of CNN. After data input, the filter traverses the data with a specific step size, that is, performs convolution operation. In this paper, the maximum pooling algorithm is adopted, that is, the maximum value in the corresponding area is extracted with a specific step size to form a new feature map.

We first convert one-dimensional coding sequence into two-dimensional, and then input it into CNN, as shown in Fig. 31.3. Two-dimensional data passes through convolution layer, pooling layer and full connection layer in turn, after which a softmax function is connected and, finally, outputs the estimated probability **p**.

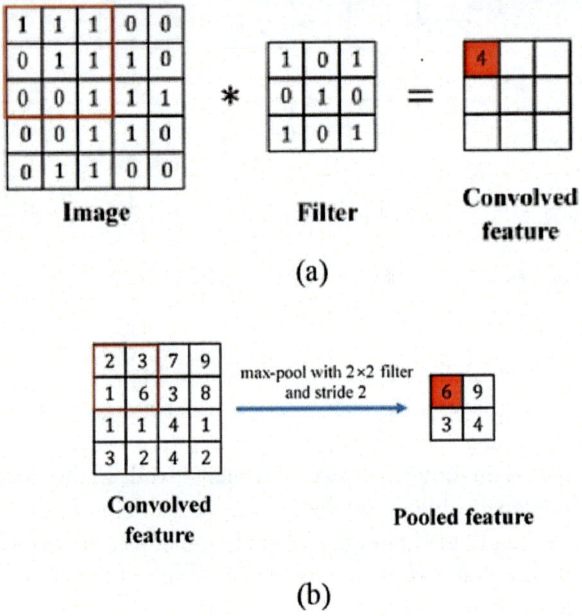


**Fig. 31.2** **a** Convolution layer with filter size of 3 × 3 and step size of 1; **b** Maximum pooling layer with filter size of 2 × 2 and step size of 2

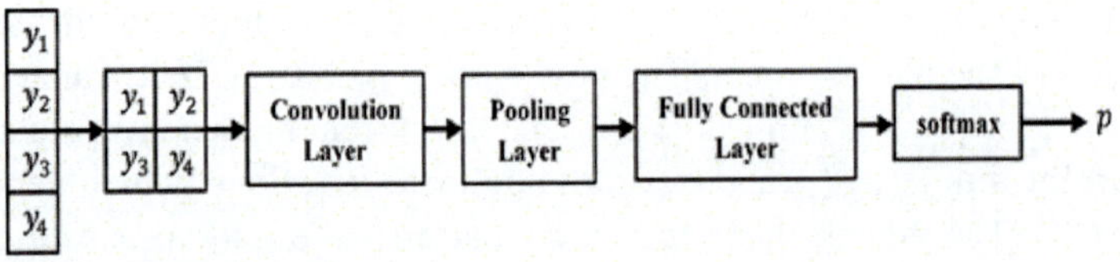


**Fig. 31.3** Structure of CNN recognizer

## 31.3.3 Training Process

Training set is a dataset of known labels, in other words, a sequence of known coding types. $\hat{\mathbf{p}}$ is one-hot coding representation of all coding type:

$$\hat{\mathbf{p}} = (C_1, C_2, \ldots C_n) \tag{31.1}$$

where $n$ is the number of encoding types. If $C_i (i = 1, 2, \ldots, n)$ is the coding type used by the sender, $C_i = 1$, and other $C_j = 0 (j \neq i)$. In the training process, we use the sequence of known coding types to find the appropriate deep neural network parameter $\theta$, which makes:

$$\theta^* = \underset{\theta}{\operatorname{argmin}}\, L(\mathbf{p}, \hat{\mathbf{p}}) \tag{31.2}$$

where $L$ is the loss function [7]. The final ideal result is that the loss value is infinitely close to 0 through training.

## 31.4 Experimental Results

In this section, we give some experiments to evaluate the performance of the recognizer based on deep learning.

In this paper, TensorFlow deep learning framework is used, and GPU is used to accelerate training. In order to use the deep learning model for training, it is necessary to generate a large amount of sample data. In this paper, SC-LDPC code and QC-LDPC code with a rate of 1/2 and lengths of 36, 72, 144, 288, 400 and 512, respectively, are generated by MATLAB, with 100,000 frames each. Then, the encoded sequence is modulated by binary phase shift keying (BPSK) and then transmitted in additive white Gaussian noise (AWGN) channel. The signal-to-noise ratio (SNR) is 0–8 dB, and the demodulated sequence constitutes the dataset. In the sample dataset, the ratio distribution of training set and test set is 8:2. The loss function adopts cross entropy loss function, and the gradient descent is Adam adaptive learning rate optimization algorithm.

Figure 31.4 depicts the changing trend of the accuracy of the two recognizers with the sequence length from 36 to 512 when SNR = 5 dB. It can be seen from the figure that the accuracy of the two LDPC codes increases with the increase of sequence length, and the performance of CNN recognizer is slightly higher than that of LSTM recognizer. Generally speaking, the recognition accuracy of the two recognizers is very high. Even when the sequence length is short, the two curves are always higher than 93%. The recognition accuracy is close to 100% when the sequence length is greater than 300. These experimental results show that our recognizer is excellent in recognizing different types of LDPC codes.

Figure 31.5 depicts the changing trend of the accuracy of the two recognizers with the SNR from 0 to 8 dB when sample length of 512. Obviously, the coding classification performance of the two algorithms increases with the increase of SNR. The performance of the two algorithms based on deep learning is almost the same.

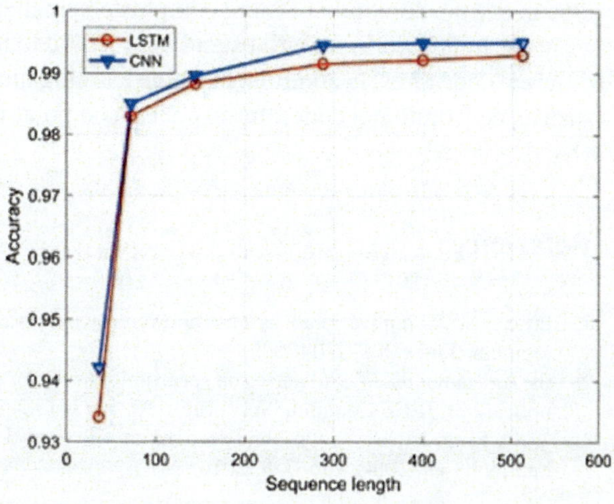


**Fig. 31.4** Identification accuracy of SC-LDPC and QC-LDPC with the sequence length from 36 to 512 when SNR = 5 dB

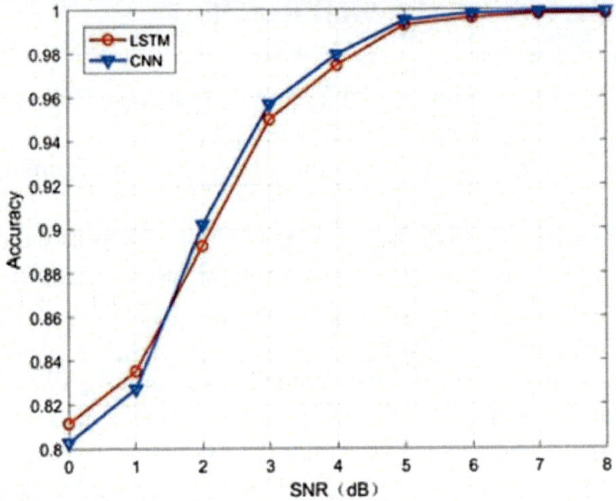


**Fig. 31.5** Identification accuracy of SC-LDPC and QC-LDPC with the SNR from 0 to 8 dB when sample length of 512

When the SNR is 0 dB, the recognition accuracy is greater than 80%, and when the SNR is greater than 5 dB, the recognition accuracy is close to 100%. The reason for this phenomenon may be that under the condition of low SNR, the neural network learns too many noise patterns in the training stage, which leads to over-fitting. In any case, the results show that our algorithm can work well in the low SNR region.

## 31.5 Conclusion

In this paper, we consider using deep learning method to realize blind recognition of channel coding. This method can effectively avoid the complex operation of extracting data features manually and save a lot of labor costs. Meanwhile, the experimental results show that this method can still effectively identify the coding type under the condition of more complex channel environment.

In the future work, we will extend our method to identify more LDPC codes with different code rates. In addition, machine learning methods and other channel coding such as convolutional codes, turbo codes, and polar codes will also be considered.

## References


1. Haykin, S.: Cognitive radio: brain-empowered wireless communications. IEEE J. Sel. Areas Commun. **23**(2), 201–220 (2005)
2. Ali, A., Hamouda, W.: Advances on spectrum sensing for cognitive radio networks: theory and applications. IEEE Commun. Surv. Tutor. **99**, 1277–1304 (2017)
3. Yu, P., Peng, H., Li, J.: On blind recognition of channel codes within a candidate set. IEEE Commun. Lett. **20**(4), 736–739 (2016)

4. Peng, S.L., Jiang, H.Y., Wang, H.X., Alwageed, H.: Modulation classification based on signal constellation diagrams and deep learning. IEEE Trans. Neural Netw. Learn. Syst. **99**, 718–727 (2018)
5. Ni, Y., Peng, S., Zhou, L., Yang, X.: Blind identification of LDPC code based on deep learning. In: 2019 6th International Conference on Dependable Systems and Their Applications (DSA), Harbin, China, pp. 460–464 (2020)
6. Qin, X., Peng, S., Yang, X., Yao, Y.: Deep learning based channel code recognition using TextCNN. In: 2019 IEEE International Symposium on Dynamic Spectrum Access Networks (DySPAN), Newark, NJ, USA, pp. 1–5 (2019)
7. Shen, B., Wu, H., Huang, C.: Blind recognition of channel codes via deep learning. In: 2019 IEEE Global Conference on Signal and Information Processing (GlobalSIP), Ottawa, Canada, pp. 1–5 (2019)

# Chapter 32
# Influence and Optimization of Levenberg–Marquardt Algorithm on Dielectric Loss Measurement in Noisy Environment

**Chao Yu, Ronghui Hu, Xiaoxiao Xu, and Suaisuai Zhao**

**Abstract** In the calculation of medium loss angle, Levenberg–Marquardt algorithm (LM algorithm for short) can improve the high-order sinusoidal fitting method in the case of no noise, and the calculation amount is large and the speed is relatively slow. However, white noise and quantization noise exist in the actual measurement of dielectric loss angle. In order to extend LM algorithm to the actual measurement of medium loss angle, this paper first studies the influence of white noise on LM algorithm in the number of iterations and calculation accuracy and then studies the change of LM algorithm after signal noise reduction through FIR narrowband bandpass filter, as well as the change of iteration times and calculation accuracy. The research shows that noise level is positively correlated with the iteration times of LM algorithm and negatively correlated with the accuracy. After noise reduction, LM algorithm can be greatly simplified, the number of iterations is reduced, and the accuracy is also improved.

## 32.1 Introduction

The dielectric loss Angle $\delta$ is the complementary angle of the phase difference between voltage and current under sinusoidal AC voltage. The tan$\delta$ is called the tangent of medium loss, which is an important index reflecting the insulation performance of high voltage electrical equipment [1]. The change of dielectric loss Angle can reflect the insulation defects such as damp exposure, deterioration or gas discharge in the insulation, so the detection of dielectric loss of electrical equipment is of great value to the safe operation of power system [2, 3].

Dielectric loss Angle is a small value, so it is greatly affected by voltage measurement. However, the voltage measurement value is easy to be distorted by frequency offset and harmonic interference [4]. In the calculation of medium loss, one of the most basic algorithms is the least square method [5], but this algorithm only considers the fundamental wave, treats the higher harmonics as noise, and thinks that the power

C. Yu (✉) · R. Hu · X. Xu · S. Zhao
Department of Electrical Engineering, Shanghai Dianji University, SDJU, Shanghai, China

 
L. C. Jain et al. (eds.), *Smart Communications, Intelligent Algorithms and Interactive Methods*, Smart Innovation, Systems and Technologies 257,
https://doi.org/10.1007/978-981-16-5164-9_32

grid frequency remains unchanged at 50 Hz. Then the improved algorithm is higher-order sine fitting algorithm [6]. The variation of high harmonics and power grid frequency is considered, but the calculated equation is a high-order nonlinear equation, which requires the use of Newton iterative algorithm, and the initial value of iteration is obtained by the discrete Fourier algorithm. Since the Hessian matrix in Newton iterative algorithm is a second-order matrix, calculation and code implementation are troublesome, and iteration takes up a lot of computing resources, so it will take a long time to get a more accurate value. Later, the improved algorithm uses the Levenberg–Marquardt algorithm (abbreviation is LM) to iterate [7], and the algorithm uses Jacobian matrix and the identity matrix with parameters to approximate the Hessian matrix. In reference 6, under the condition of no noise, the LM algorithm is superior to the Newton iteration method in terms of algorithm simplicity and final fitting results and speed. But in practical engineering, there will be white noise, impulse noise and quantization noise [8, 9], and white noise is the most common noise that has the greatest impact on the results. Therefore, in order to extend LM algorithm to the actual measurement of dielectric loss, the change of the accuracy and iteration times of LM algorithm caused by white noise, as well as the improvement of the algorithm in this case, is the direction considered in this paper.

## 32.2 High-Order Sinusoidal Fitting Method Is Used to Calculate the Dielectric Loss

The fast Fourier algorithm can obtain the amplitude and phase of each frequency in the case of quasi-synchronous sampling [10], but the power grid frequency is not a fixed value and fluctuates to a certain extent. Therefore, sampling cannot be carried out synchronously, and spectrum aliasing will occur. Although the higher-order sinusoidal fitting is also based on the Principle of Fourier transform, it takes frequency as the unknown quantity and uses the method of harmonic weighting to fit the voltage and current signal. Decompose the measured voltage and current signal into.

In the Fig. 32.1, the measured waveform is decomposed into DC component, fundamental wave and odd harmonic component (even harmonic amplitude is very small and can be ignored), and the fundamental wave is decomposed into sine and cosine components again for the convenience of calculation. The power grid

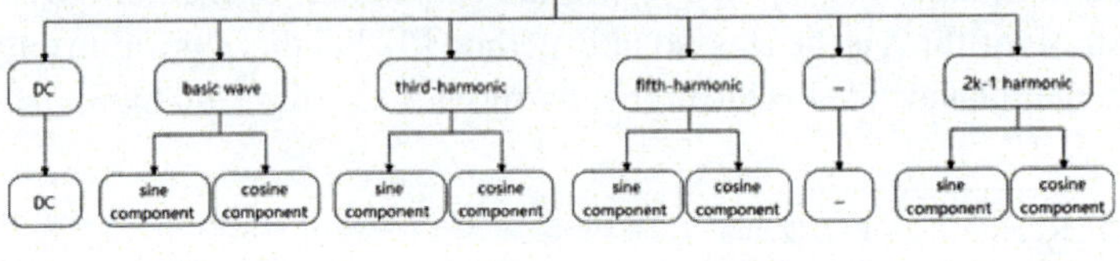


**Fig. 32.1** Original waveform decomposition

frequency is a fluctuating value, but it can be assumed that the power grid frequency does not change in a very short time. This paper assumes that the power grid frequency remains unchanged in a power frequency cycle. The current and voltage signals can be represented by the following mathematical model:

$$I(t_n) = A_0 + \sum_{k=1}^{N} (A_{K1}\sin((2k-1)2\pi f t_n) + A_{K2}\cos((2k-1)2\pi f t_n)) \quad (32.1)$$

$$U(t_n) = B_0 + \sum_{k=1}^{N} (B_{K1}\sin((2k-1)2\pi f t_n) + b_{K2}\cos((2k-1)2\pi f t_n)) \quad (32.2)$$

In this equation, $A_0$, $B_0$ is the DC component; $A_{k1}$, $A_{k2}$, $A_{k3}$, and $A_{k4}$ are the amplitude of the sine and cosine components of the $(2k-1)$ harmonic, respectively, $t_n$ is the $n$th time point. The two mathematical models are used to fit the actual voltage and current signals. The fitting error can be expressed as:

$$e_i(n) = I(t_n) - i(t_n) \quad (32.3)$$

$$e_u(n) = U(t_n) - u(t_n) \quad (32.4)$$

The squared sum representation of the error is as follows:

$$E_i = \sum_{n=0}^{M-1} (I(t_n) - i(t_n))^2 \quad (32.5)$$

$$E_u = \sum_{n=0}^{M-1} (U(t_n) - u(t_n))^2 \quad (32.6)$$

In this equation, $i(t_n)$ and $u(t_n)$ are the current and voltage series obtained in practice. In order to minimize the sum of squares of errors, an appropriate nonlinear optimization algorithm can be used to achieve this.

After each parameter is determined, the tangent of the medium loss Angle can be expressed as:

$$\tan\delta = \frac{A_{12}B_{12} + A_{11}B_{11}}{A_{11}B_{12} - A_{12}B_{11}} \quad (32.7)$$

In this equation, $A_{11}$, $A_{12}$, $B_{11}$, and $B_{12}$ are the amplitude of sine and cosine components of the fundamental wave in the voltage and current signal.

## 32.3 Levenberg–Marquardt Algorithm

The LM algorithm is a numerical solution that provides nonlinear minimization. By modifying parameters during execution, the proposed algorithm achieves the advantages of both Gauss–Newton algorithm and gradient descent algorithm and improves the shortcomings of both algorithms.

When LM algorithm is used, the recursive equation is:

$$\boldsymbol{X}(m+1) = \boldsymbol{X}(m) - \boldsymbol{H}_i{}^{-1}\boldsymbol{J}_i(m)^T\boldsymbol{e}_i(m) \tag{32.8}$$

$$\boldsymbol{Y}(m+1) = \boldsymbol{Y}(m) - \boldsymbol{H}_u{}^{-1}\boldsymbol{J}_u(m)^T\boldsymbol{e}_u(m) \tag{32.9}$$

In this equation, $X$ and $Y$ are variables of nonlinear functions of multiple variables, $X(m)$ and $Y(m)$ are the results obtained by the $m$th iteration, $X(m+1)$ and $Y(m+1)$ are the result of the No $(m+1)$ iteration, $H_i$ and $H_u$ is the Hessian matrix, $J_i$ and $J_u$ is the Jacobian matrix, $e_i(m)$ and $e_u(m)$ are the corresponding error column vector, and m is the number of iterations.

To reduce repetition, take current as an example to expand the recursive equation, where the expression of $H$ matrix is:

$$\boldsymbol{H}_i(m) \approx \boldsymbol{J}_i(m)^T\boldsymbol{J}_i(m) + \lambda\boldsymbol{I} \tag{32.10}$$

In the equation, $\boldsymbol{I}$ is the $2N+2$ orders identity matrix, $\lambda$ is an adjustable parameter, and $\boldsymbol{J}_i(\boldsymbol{n})$ is Jacobian matrix. This matrix is obtained by taking the first-order derivative from $\boldsymbol{I}$. The first-order derivative at the $n$th point can be expressed as:

$$\boldsymbol{J}_{i_n} = \frac{\partial I(t_n)}{\partial \boldsymbol{A}} \tag{32.11}$$

In this equation, **A** is the $2+2N$ dimensional variable vector, $A = \{A_0, f. A_{11}, A_{12}, \ldots, A_{k1}, A_{k2}\}$, $A_{k1}$, and $A_{k2}$ are the amplitude of the $(2k-1)$ harmonic sine and cosine component. The partial derivative of each variable is obtained as follows:

$$J_{i_{n,1}} = \frac{\partial I(t_n)}{\partial A_0} = 0 \tag{32.12}$$

$$J_{i_{n,2}} = \frac{\partial I(t_n)}{\partial f} = \sum_{k=1}^{n}(A_{k1}(2k-1)2\pi t_n \cos((2k-1)2\pi f t_n) - A_{k2}(2k-1)$$
$$2\pi t_n \sin((2k-1)2\pi f t_n)) \tag{32.13}$$

$$J_{i_{n,2k+1}} = \frac{\partial I(t_n)}{\partial A_{k1}} = \sin((2k-1)2\pi f t_n) \tag{32.14}$$

$$J_{i_{n,2k+2}} = \frac{\partial I(t_n)}{\partial A_{k1}} = \cos((2k-1)2\pi f t_n) \tag{32.15}$$

If the data sampling is out of sync, the results obtained by directly calculating the signal parameters using FFT are not accurate. However, due to the small fluctuation of the power grid frequency, it is expected that the obtained results are not different from the actual signal parameters. Therefore, the initial value of iteration except the frequency (when $m = 0$) can be obtained by the current signal changing through fast Fourier, and the frequency is set as 50 by default. Taking these values as initial values can reduce the number of iterations and speed up the calculation. The flowchart of high-order sinusoidal fitting medium loss Angle based on LM algorithm is as follows (Fig. 32.2).

To analyze the influence of white noise on LM algorithm, the frequency was fixed at 50 Hz, and only five or less harmonics were considered. The amplitude parameters

**Fig. 32.2** LM algorithm flow chart

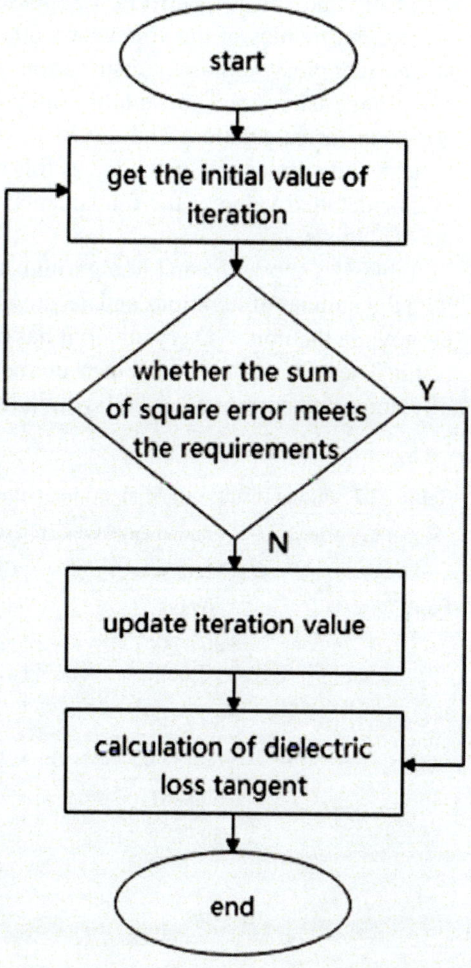

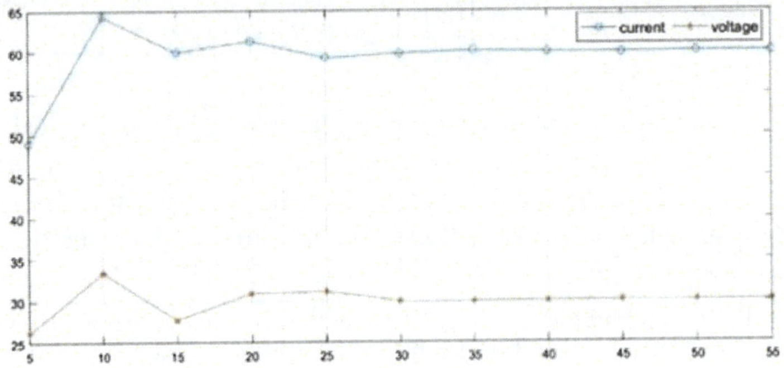


**Fig. 32.3** Fundamental phase diagram of voltage and current under noise

of voltage and current signal are consistent with literature 6. The phase angles of the fundamental current wave, the third harmonic wave, and the fifth harmonic wave are 30°, 150°, and 75°, respectively. The phase angles of the fundamental, the third, and the fifth harmonics of the voltage are 60, 148, and 78°, respectively. Is the voltage and current plus two normal distribution random noises with the same SNR and the SNR range is: 5–55. The sampling frequency is 10,000, and 200 sampling points are taken for calculation. The results of the fundamental wave phase Angle of the voltage and current calculated are as follows (Fig. 32.3).

The detailed value of the fundamental wave phase and the number of iterations are as follows:

Table 32.1 shows that in LM algorithm, the amount of noise is positively correlated with the number of iterations and negatively correlated with the accuracy. Moreover, even when the noise is very small, it has a great impact on the accuracy of the final result. Therefore, in the actual measurement of medium loss, the sampling sequence of voltage and current needs to be digitally filtered first.

**Table 32.1** Fitting results of LM algorithm with different SNR

| Signal to noise ratio | Fundamental wave phase angle | | The number of iterations | |
|---|---|---|---|---|
| | Voltage | Current | Voltage | Current |
| No noise | 60.0000 | 30.0000 | 0 | 0 |
| 5 | 49.0255 | 26.2268 | 6 | 6 |
| 15 | 59.9795 | 27.7562 | 3 | 4 |
| 25 | 59.2745 | 31.1238 | 3 | 3 |
| 35 | 60.1334 | 29.9385 | 2 | 2 |
| 45 | 59.8481 | 30.0352 | 2 | 2 |
| 55 | 59.9899 | 30.073 | 2 | 2 |

## 32.4 FIR Narrowband Bandpass Filter

In the online measurement of dielectric loss, the required frequency is the frequency near the power frequency. If there is a narrowband-pass filter of 49–51 Hz, the high-order harmonics and most of the noise can be removed. In the calculation, the high-order harmonics need not be taken into account, and the reduction of noise can also reduce the number of iterations.

The calculation of dielectric loss Angle is to find the complementary Angle of voltage and current phase difference, and the filter must be strictly linear phase to ensure that the calculated results will not generate new errors due to the filter. FIR filter is a full zero point system, that is, the $Z$ transfer function at the $Z$ complex plane poles are all at $Z = 0$, which makes the FIR filter have very good linear phase.

FIR filter design methods have window function design method and frequency sampling design method. The frequency sampling design method is designed from the frequency domain, which is especially suitable for narrowband bandpass filters. Therefore, this paper uses the frequency sampling design method to design narrowband bandpass filters of 49–51 Hz.

In the frequency range, suppose there is a target frequency response $H_d(e^{jw})$, which is a continuous function about $w$, and now sample $N$ points at equal intervals on the unit circle of $e^{jw} = 1$, and get $H_d(k)$, and this is the actual frequency response:

$$H_d(k) = H_d\left(e^{j\omega}\right)|_{\omega_k=2\pi k/N} = H_d\left(e^{j2\pi k/N}\right) \tag{32.16}$$

$H_d(k)$ is transport by the inverse Fourier transform to the time domain, and the required impulse response is obtained:

$$h(n) = \frac{1}{N}\sum_{k=0}^{N-1} H_d(k)e^{\frac{j2\pi k}{N}}, n = 0, 1, 2, \ldots, N-1 \tag{32.17}$$

When giving the required frequency response $H_d(k)$ in the time domain, generally only the value of $0 \sim \pi$ will be given, so the value of $\pi \sim 2\pi$ needs to be filled up according to the linear phase condition of FIR filter before doing the Inverse Fourier transform. The ideal frequency response is defined as:

$$H\left(e^{j\omega}\right) = H(\omega)e^{j\theta(\omega)} \tag{32.18}$$

This expression represents the multiplication of amplitude response and phase response. To satisfy the linear phase condition, it can be expressed as:

$$\begin{cases} H(N-k) = H(k) \\ \theta(N-k) = \frac{N-1}{N}\pi k \end{cases}, k = 1, 2, \ldots, \frac{N-1}{2} \tag{32.19}$$

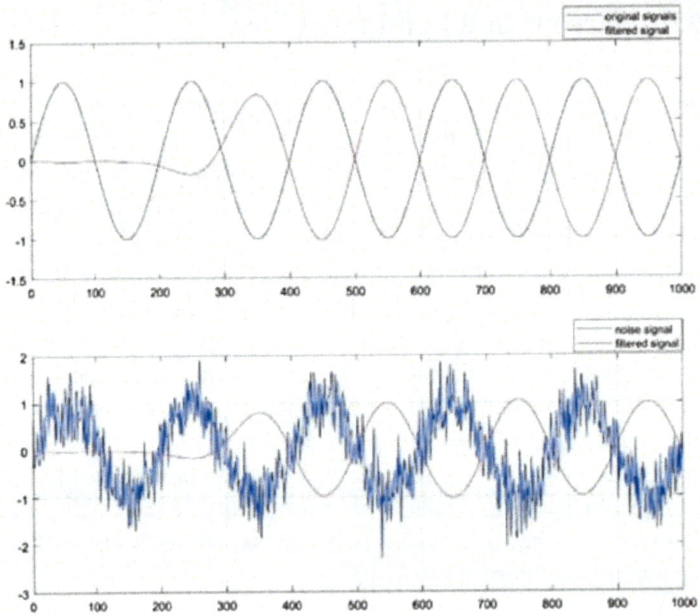


**Fig. 32.4** Filter effect chart

In this equation, $\theta(k) = -\frac{N-1}{N}\pi k(k) = -(N-1)k\pi/N$, when $N$ is even, $H(N/2) = 0$.

The order of FIR filter is large enough to make the filtered waveform smooth enough, and the filter has better linear phase when the order is odd, so the selected order is close to 599 of three periods, that is, 599 sampling points on the unit circle. The corresponding sampling points of 49–51 Hz are retained, and the remaining sampling points are set to zero. A standard sine wave with frequency of 50 Hz and a 50 Hz sine wave with noise are put into the filter, and the results are as follows (Fig. 32.4).

As can be seen in the figure, this filter will not distort the original waveform, and the filtering effect can also meet the requirements. After filtering, the first part of waveform is different from the original waveform. This is because the FIR filter carries out convolution operation with signals in the form of window. Only after the whole window enters can the waveform be completely restored. Since the order of this filter is 599, the original waveform is completely restored from the 600th point. After calculation, the phase of the filtered waveform is shifted back by 178.2 degrees compared with the original waveform. In this paper, the filtered phase results are obtained after phase compensation.

The simulation data of voltage and current is the same as the previous one, but 1000 sampling points are taken to enter the filter. After filtering, take 200 points from the 600th to 799 and enter into LM algorithm calculation. Because this digital filter can filter out all the higher harmonics, the variables are reduced to only $A0$, $A11$, $A12$, and $F$.

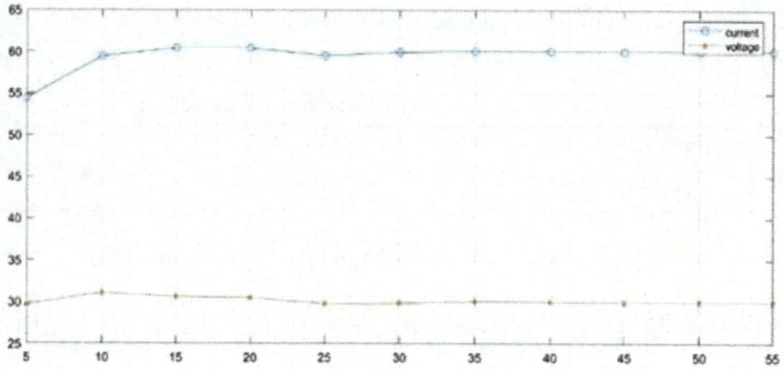


**Fig. 32.5** Fundamental phase diagram of voltage and current after filtering

The voltage and current signal are the same as the previous section. The phase Angle of the fundamental wave of the voltage and current calculated is as follows (Fig. 32.5).

The detailed value of the fundamental wave phase and the number of iterations are as follows.

The comparison between Tables 32.1 and 32.2 shows that the FIR narrowband bandpass filter can reduce the iteration times of LM algorithm when the noise is large and improves the accuracy of results when the noise is small.

When the SNR is 55 and the frequency fluctuates between 49.5–50.5, with a filter, the results are as follows.

When the SNR is 55, the frequency fluctuates between 49.5–50.5, and there is no filter, and the results are as follows.

It can be seen from Tables 32.3 and 32.4 that, during the non-integral period sampling, the phase of the obtained signal will deviate from the actual value due to the fact that the point is fixed from the 600th point after filtering. However, the voltage and current offsets are basically the same, so the effect on the results can be ignored. So when noise and frequency fluctuations are present, and FIR bandpass filters can also improve the accuracy of the final results.

**Table 32.2** Fitting results of LM algorithm with different SNR after denoising

| Signal to noise ratio | Fundamental wave phase angle | | The number of iterations | |
|---|---|---|---|---|
| | Voltage | Current | Voltage | Current |
| There is no noise | 60.0000 | 30.0000 | 0 | 0 |
| 5 | 54.4482 | 29.6957 | 3 | 3 |
| 15 | 60.4249 | 30.6393 | 3 | 3 |
| 25 | 59.5213 | 29.8220 | 3 | 3 |
| 35 | 60.0593 | 30.1198 | 1 | 1 |
| 45 | 59.9796 | 30.0116 | 1 | 1 |
| 55 | 60.0124 | 30.0047 | 1 | 1 |

**Table 32.3** Fitting results of LM algorithm with different frequencies after noise reduction

| Frequency | Fundamental wave phase angle | | The number of iterations | |
|---|---|---|---|---|
| | Voltage | Current | Voltage | Current |
| 49.5 | 54.5787 | 24.5723 | 3 | 3 |
| 49.7 | 56.7539 | 26.7527 | 3 | 3 |
| 50 | 59.9991 | 30.0016 | 1 | 1 |
| 50.2 | 62.1674 | 32.1630 | 3 | 3 |
| 50.5 | 65.4133 | 35.4185 | 3 | 3 |

**Table 32.4** Fitting results of LM algorithm at different frequencies

| Frequency | Fundamental wave phase angle | | The number of iterations | |
|---|---|---|---|---|
| | Voltage | Current | Voltage | Current |
| 49.5 | 60.0266 | 29.9899 | 3 | 3 |
| 49.7 | 59.9797 | 29.9766 | 3 | 3 |
| 50 | 59.9965 | 30.0143 | 2 | 2 |
| 50.2 | 59.9819 | 30.0212 | 3 | 3 |
| 50.5 | 60.0053 | 29.9793 | 3 | 3 |

## 32.5 Conclusion

The high-order sinusoidal fitting method can obtain the intermediate loss value in the case of unknown grid frequency, but LM algorithm is needed to solve the problem. However, in the actual measurement of dielectric loss, LM algorithm has a great error due to the presence of white noise, and the number of iterations also increases significantly. Therefore, for LM algorithm, noise reduction before calculation is necessary. FIR narrowband bandpass filter can filter out most of the noise and higher harmonics, greatly simplifying LM algorithm, algorithm iteration times in the noise more significantly reduced, and accuracy is also significantly improved. In addition, FIR filter can also improve the accuracy of results when frequency fluctuation and noise exist simultaneously.

## References

1. Liao, J.Q., Liang, G.X.: A method for measuring dielectric loss combining complex trigonometric function and Hilbert transform. Insul. Mater. **45**(3), 69–72 (2012)
2. Tian, C.F.: Correct measurement of capacitance capacity and dielectric loss of transformer capacitive bushing. Electr. Autom. Equip. **30**(4), 147–149 (2010)
3. Peng, X., Chen, H., Xia, G.L. et al.: Field high voltage dielectric loss measurement of shunt capacitors with 500 kV circuit breakers. High Voltage Technol.**37**(10), 2378–2384 (2011)

4. Zhang, W., Lv, Z.C., Deng, Y.R.: Application of fundamental wave frequency correction method in measuring dielectric loss. Insul. Mater. **4**, 77–80 (2016)
5. Zhao, X.S., Tan, K.X., Zhu, D.H., et al.: Digital measurement of dielectric loss angle. J. China Univ. Sci. Technol. (Nat. Sci.) **36**(9), 51–56 (1996)
6. Wang, W.L., Li, F.Q., Tan, K.X.: High order sinusoidal fitting method for measuring dielectric loss Angle. J. Tsinghua Univ. (Nat. Sci. Ed.) **41**(9), 5–8 (2001)
7. Xu, Z.N., Feng, G.Y., Lu, F.C., et al.: Comparison between Newton method and Levenberg-Marquardt algorithm and its application in mesoloss Angle measurement. J. North China Electr. Power Univ. **33**(1), 15–19 (2006)
8. Chen, L.A.: Modified Levenberg-Marquardt method with line search for nonlinear equations. SCI Comput. Optim. Appl. **3**, 753–779 (2016)
9. Xu, Z.N.: Influence of noise on the tangent calculation results of dielectric loss Angle. Electr. Power Autom. Equip. **33**(11), 69–74 (2013)
10. Wang, M., Vandermaar, A.J., Srivastava, K.d.: Review of condition assessment of power insulation in service. IEEE Electr. Insul. Mag. **18**(6), 12–25 (2002)

# Chapter 33
# Research on Artificial Intelligence Optimization Based on Genetic Algorithm

**Jingyi Liu**

**Abstract** As an important force for national construction and development, science and technology are also a major factor affecting global economic development. Especially in the rapid development of artificial intelligence technology, genetic algorithm, as one of its optimization methods, has played a positive role in the field of innovation and technological innovation, and has changed people's life and work in a certain sense. Therefore, on the basis of understanding artificial intelligence and genetic algorithm, this paper further studies how to carry out artificial intelligence optimization based on genetic algorithm.

## 33.1 Artificial Intelligence

Artificial intelligence, also known as artificial intelligence, was first proposed by Hugo de Gareth, an American scientist. It refers to an artificial system that can sense, understand, make decisions and execute decisions as well as humans. Artificial intelligence can be understood in two aspects: artificial intelligence is human processing and manufacturing, while intelligence includes consciousness, thinking and self. In today's social environment, the realization of intelligence requires three elements: first, big data; second, the Internet; third, cloud computing [1].

## 33.2 Genetic Algorithm

To put it simply, genetic algorithm refers to the computational model constructed by means of genetics and natural selection in Darwinian biological evolution, which can search for the optimal solution. Generally speaking, the steps of genetic algorithm are mainly divided into the following: first, estimate the adaptation degree of each chromosome; second, choose two individuals from a population and treat them as

J. Liu (✉)
Qingdao University of Science and Technology, Qingdao, China
e-mail: T_Tlll@163.com

 
L. C. Jain et al. (eds.), *Smart Communications, Intelligent Algorithms and Interactive Methods*, Smart Innovation, Systems and Technologies 257,
https://doi.org/10.1007/978-981-16-5164-9_33

parents. Third, obtain the parents' chromosomes and obtain the offspring in the cross experiment; fourth, the chromosome variation of the progeny was treated. Fifth, repeat steps 2, 3, and 4 until a new species is created. Among them, the basic process box is shown in Fig. 33.1, and the pseudo-code of the simple operation process of genetic algorithm is shown in Fig. 33.2. In the cycle operation, researchers can find satisfactory solutions to various problems, and the technical terms involved are shown in Table 33.1 [2–4].

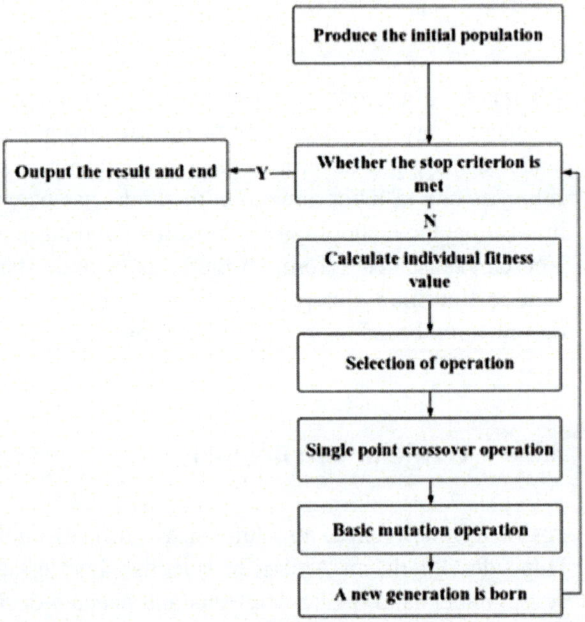


**Fig. 33.1** Basic flow diagram

```
choose an intial popualation
For each h in population,compute Fitness(h)W hile(max Fitness(h) <Fitnessthareshold)do selection
do crossaverdo mutation
update population
For cach h in population,compute Fitness(h)Return best Fitness
```

**Fig. 33.2** Simple operation flow pseudo-code

**Table 33.1** Description of biogenetic terms in genetic algorithm

| | The term | explain |
|---|---|---|
| 1 | Chromosome | Also known as genotype individuals, having a certain number of individuals forms a population, and the number of individuals in a population is the size of the population |
| 2 | Gene | Used to represent individual characteristics, such as a string $S = 1011$, where 1, 0, 1, and 1 are genes |
| 3 | Gene loci | Represents the position of a gene in the string, also referred to as the gene position, for example, the position of $S = 1011$, 0 is 2 |
| 4 | The eigenvalue | If integers are represented by strings, the eigen value of a gene has the same weight as the binary number |
| 5 | Fitness | It refers to the degree to which an individual ADAPTS to the environment. To demonstrate the fitness of a chromosome, a function of measure is introduced for each chromosome in question, also known as the fitness function |

## 33.3 Research on Artificial Intelligence Optimization Based on Genetic Algorithm

### *33.3.1 Robocode*

In July 2001, the Robocode proposed by IBM of the United States was a tank robot combat simulation engine, which had the following advantages: first, it was based on Java development and had object-oriented concepts such as inheritance; second, it can quickly test the genetic algorithm designed by itself through competition, which is interesting and competitive to some extent. Third, the community has 4000 robots to choose from, and can simulate the real genetic environment, more have a lot of code for the development of reference; fourth, Robocode itself is open source software, and designers can directly add new genetic features to promote heredity [5–7].

### *33.3.2 Optimization of Default Strategies*

There are many genetic algorithms that can be selected for robot optimization, such as genetic movement, preset strategy, etc. The following is to analyze how to scientifically optimize Robocode by taking preset strategies as an example. Since the basic operations of such robots are scanning, moving, shooting, etc., the specific code is as follows: radar scanning strategy is radar-strategy (RS), moving strategy is move-strategy (MS), bullet energy is bullet-power-strategy (BPS), etc.

Typically, robots like Robocode are programmed with four strategies: move, aim, radar, shoot, and if something happens, a behavioral strategy for that event is triggered. Because each strategy include different forms of processing, and can be triggered by genetic algorithm (GA), there may be about 4224 kinds of possibilities, so involved in mobile function also has A lot of kinds, like set A head and ahead can let robot continues to move forward A distance, and the set Back and the Back is backward, Part of the code as shown below, the following:

```
Random:
switch(Math.randlom0*2)[
case 0;sen TurnRight(Math.randomy*90):break;
casel:
set'TurnLeft(Math.random0*90); break;}
exeeute(;
Linear:
aheacl(200):setBaek(200);Aheacl(200): setBaek(200);
Circular:
setTurnRighn 1000; setMax Nelocity(4):ahead( 1000):
SETTurnRighn 100; SETMAX Nelcity(): be in the lead; (ake the) lead; in the
highest fight; keep ah.
ead; on top (1000):
```

By using genetic algorithm to adjust the Robocode motion position, we can start from two aspects: first, master the position of the robot, make clear our own speed, and know the position, relative angle and speed of the opponent; second, master the robot's motion position and know the opponent's position, relative angle and speed. Second, the orientation, bullet position and size parameters of all robots on the field. Input the above information into the system, which can help the robot to calculate the accurate angle and distance when it moves. It should be noted that the program code should meet the following requirements: first, use the function to get the specific value of fitness; Second, genetic algorithm is used to operate the robot when it moves. Third, in order to obtain the genetic offspring in the exercise, you must not manually enter it after the end [8, 9].

Because the genetic algorithm is to achieve the movement must choose two categories, one is GA, and the other is move pattern. The former is relatively simple, mainly taking values and defining groups, and then reading and writing files based on the definition and data manipulation. The common data include the probability of crossover and mutation, population size and the weight coefficient of the equation. The latter can complete the movement according to cross variation, and the output content will be stored in the vector function. Since this function contains a set of real numbers, it can be used not only to calculate the $x$ and $y$ axes, but also to compute precise values such as distances and angles.

In move pattern, for example, Grad, Input, and Rang represent the order, the input number and the weighted range of the function, respectively. During cross movement, a new feature code will appear in each operation, which is an array of variables between 0 and 1, and operations can be carried out according to the weighted value in the function vector. As the operation of this position is relatively simple, any value within the weighted range is randomly selected to replace the array of variables and stored in the movement mode. The specific variation code is shown in below:

```
protected void mutate0{
Reights X[inty (Math.random 0"weightsX.length)] = Math.randomk); range*2-
range;
weightsY(int) (Math.random (*weight.Xlenghy] = Math.randlomJ"range2- -
range;
```

After the completion of the above genetic algorithm optimization work, if the Robocode does not store the values in the working state, then the system will randomly acquire the new movement mode according to the input strategy, otherwise the system will automatically load and store the values. In the initial stage, the corresponding fitness will be proposed for each mobile mode, and the system will accurately number and implement the following steps: first, the mobile mode will be graded according to the specific value of fitness; second, elite, crossover and mutation operation should be carried out according to the requirements. Third, store the weighted value; fourth, redo the genetic algorithm. In the continuous adjustment of the system, researchers can get the best adaptive value, and thus, propose a more valuable optimization scheme.

## 33.4 Conclusion

To sum up, compared with the previous optimization forms, genetic algorithm consumes less time, has high accuracy, is simple to operate, has strong expansibility and inspiration, and can integrate with other algorithms faster. Therefore, in the artificial intelligence in the process of rapid development, in the face of more and more high-technical requirements, while GA has made certain achievements, but in order to better solve the problem of artificial intelligence optimization, dealing with intelligent perception, and cognitive function in the system, must want to continue to strengthen research efforts, and combine with practice case analysis, only in this way can quickly achieve expected goals. And in future development, I believe that through our continuous exploration, we will be able to make new breakthroughs in the manufacture and application of artificial intelligence technology.

## References

1. Lin, M.T., Zhao, Y.P.: Artificial intelligence-empowered resource management for future wireless communications: a survey. China Commun. **17**(03), 58–77 (2020)
2. Patle, B.K., Ganesh Babu, L., Pandey, A., Parhi, D.R.K., Jagadeesh, A.: A review: on path planning strategies for navigation of mobile robot. Def. Technol. **15**(04), 582–606 (2019)
3. Lin, C., Li, S., Wu, K., Liu, T.: Intelligent assessment method of air-defense & anti-missile command model based on genetic algorithm optimization. In: IOP Conference Series: Materials Science and Engineering, vol. 435(1), pp. 66–71 (2018)
4. Zhao, X., Li, Y., Boonen, P.: Intelligent optimization algorithm of non-convex function based on genetic algorithm. J. Intell. Fuzzy Syst. **35**(4), 15–159 (2018)
5. Azadeh, G., Hossein, B., Isa, E., Majid, M., Bahram, G., Saeed Reza, K.: Uncertainty analysis of intelligent model of hybrid genetic algorithm and particle swarm optimization with ANFIS to predict threshold bank profile shape based on digital laser approach sensing. Measurement 121 (2018)
6. Li, A., Zhao, W.Z., Li, S.M., Qiu, X.Y., Wang, X.B.: Research on the motion trajectory optimization method based on the improved genetic algorithm for an intelligent vehicle. Proc. Inst. Mech. Eng. **230**(13) (2016)
7. Wang, X.R., Zhou, F., Wang, X.W., Guo, J.: Intelligent optimization of the structure of the large section highway tunnel based on improved immune genetic algorithm. Chem. Eng. Trans. (CET J.) 59 (2017)
8. Li, A.J., Zhao, W.Z., Li, S.M., Qiu, X.Y., Wang, X.B.: Research on the motion trajectory optimization method based on the improved genetic algorithm for an intelligent vehicle. Proc. Inst. Mech. Eng. Part D: J. Automob. Eng. **230**(13), 99–105 (2016)
9. Zhu, M., Su, B., Ning, G.M.: Intelligent optimization of urban real-time traffic route based on DNA non-dominated sorting genetic algorithm. Appl. Mech. Mater. 3744 (2015)

# Chapter 34
# Multi-channel Scheduling for Intelligent Information Collection

**Xingchuan Bao, Jianhua Yang, Jian Yang, Lin Peng, and Min Xu**

**Abstract** In order to effectively improve the service quality of intelligent information collection based on wireless sensor networks in power emergency scenarios, further enhance the real-time data in intelligent information collection and reduce delay, a priority-based multi-channel scheduling strategy is proposed. First, based on real-time channel link state information, a link routing algorithm based on minimum hop spanning tree (LB-MHST) is proposed to overcome radio frequency interference and ensure the service quality of the smart grid. Then, according to the different delay requirements of different data packets in intelligent information collection, data priority transmission is considered, which effectively improves the data transmission efficiency of the sensing node and further meets the Quality of Service (QoS) performance requirements in intelligent information collection. Experimental results show that compared to the minimum hop spanning tree (MHST) algorithm, especially in multi-channel delay-aware transmission, the proposed algorithm improves the delay performance of high-load traffic in the case of single channel, 8 and 16 channels, respectively, increased by 12%, 15.2%, and 18%.

## 34.1 Introduction

In order to ensure the QoS [1] in smart grid applications, in terms of communication interference, the influence of radio frequency interference [2] is often overcome through multi-channel communication, which not only ensures multi-channel communication, but also improves the service quality in smart grid. In terms of latency, parallel transmission is adopted to improve the overall performance of the

X. Bao (✉) · L. Peng · M. Xu
Global Energy Interconnection Research Institute, Nanjing 210003, Jiangsu, China

X. Bao
State Grid Laboratory of Electric Power Communication Network Technology, Nanjing 210003, China

J. Yang · J. Yang
State Grid ZheJiang Electric Power Co.,Ltd, Hangzhou 310000, Zhejiang, China

 
L. C. Jain et al. (eds.), *Smart Communications, Intelligent Algorithms and Interactive Methods*, Smart Innovation, Systems and Technologies 257,
https://doi.org/10.1007/978-981-16-5164-9_34

network [3]. But with the development of smart grid, the grid application is becoming more and more high to the requirement of service quality, make the most of the communication technology (as shown in Table 34.1) cannot fully guarantee required in the application of grid service quality, which requires a reliable and efficient communication architecture for smart grid applications in future to provide satisfactory service [4]. However, different communication architectures have different communication delay requirements [5]. In addition, the network traffic load and data type of data communication between communication devices are constantly changing [6], making time a key factor for data collection of smart grid.

Intelligent channel allocation strategy is the most effective way to counter channel interference. Musham et al. [7] proposed the use of a near-optimal algorithm for grid WMNs to mitigate the impact of channel interference in multi-channel transmission. Suryawanshi et al. [8] proposed to construct a multi-channel communication model against data loss caused by excessive channel interference load and recover data lost due to channel interference as far as possible. Jayalakshmi et al. [9] proposed a routing tree based on the channel information collection of network nodes for channel allocation, so as to combat interference more effectively.

However, the sensor network QoS performance is not only affected by rf interference, but also by the network topology. In the multi-channel allocation strategies against channel interference, the delay requirements of different packets are also different [10]. To further satisfy the QoS requirement in the application of information intelligent acquisition and awareness, this paper proposes a link routing algorithm based on the link grid minimum hop spanning tree (LG-MHST) for data priority channel scheduling strategy on the basis of ensuring channel interference. In addition, the priority of data is fully considered during the channel allocation, which not only guarantees the good anti-interference performance of intelligent information collection, but also reduces the transmission delay of intelligent information collection, and further comprehensively meets the QoS requirements of intelligent information collection for data transmission.

**Table 34.1** Several existing QoS aware routing protocol in wireless sensor network

| Protocol | Delay | Throughput | Reliability |
|---|---|---|---|
| RTLD [10] | No | Yes | Yes |
| RPL [10] | Yes | Yes | Yes |
| BIOSARP [11] | Yes | No | No |
| RRR [11] | Yes | Yes | Yes |
| OQAP [11] | No | No | No |
| MRL-CC [12] | No | No | No |
| DARA [12] | Yes | No | Yes |

## 34.2 Communication Model of the Main Station for Intelligent Information Collection

According to research [11], the above model can be used in a radio propagation environment with interference, such as a state-aware environment for intelligent information collection. The relevant parameters of the model are shown in Table 34.2. In this model, the path fading from the transmitter $d$ is:

$$PL_{d_0 \to d} = PL(d_0) + 10\eta \log_{10}\left(\frac{d}{d_0} + X_\sigma\right) \tag{34.1}$$

among them.

| | |
|---|---|
| $PL_{d_0 \to d}$ | is the path fading at a distance d from the transmitter, unit is dBm. |
| $PL(d_0)$ | As reference distance $d_0$ Path fading. |
| $\eta$ | Path fading coefficient. |
| $X_\sigma$ | is Gaussian random variable with mean zero and standard deviation σ. |

This paper is based on the good anti-interference performance of the receiver's channel scheduling (RBCA) [12] algorithm, combined with WSN's own network characteristics, in the later time gap allocation. The data priority is considered in parallel transmission scheduling with multi-branch, and finally a scheduling algorithm that divides channels according to data priority is proposed, that is, a link routing algorithm based on the minimum hop spanning tree. The purpose of reducing the delay of intelligent information collection and transmission is achieved, which further satisfies the QoS requirements of intelligent information collection for data transmission in complex environments.

The LB-MHST algorithm takes the sink node in the WSN as the root node, so that the link cost of each sub tree connected to the sink node is less than the present value. Then, the channel scheduling algorithm based on data priority is shown in Fig. 34.1. Build a routing tree, then use the multi-channel scheduling algorithm based on time division multiple access (TDMA) to calculate the minimum scheduling length, and implement channel allocation according to the radio frequency (RF) status between nodes. That is, if the SINR between nodes exceeds a predetermined threshold, channels can be allocated to these nodes.

**Table 34.2** Logarithmic normal shadow model parameters in smart grid environment

| Transmission environment | 500 kV ground transmission (no unit) |
|---|---|
| Path fading ($\eta$) | 2.42 |
| Shadow variance ($X_\sigma$) | 3.12 |

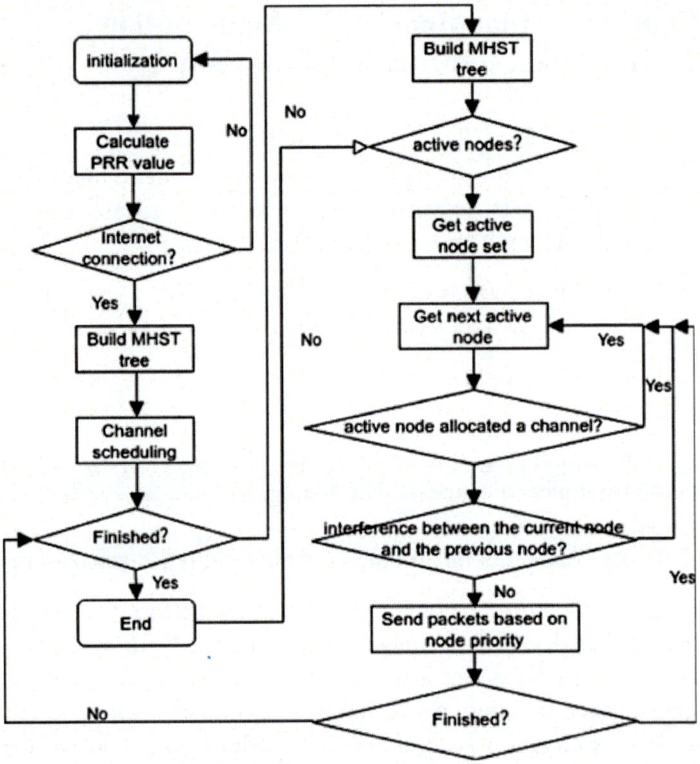


Fig. 34.1 Data priority channel scheduling algorithm

## 34.3 Algorithm

First, based on the real-time channel link state information, based on the RBCA algorithm, an improved channel allocation algorithm is proposed. As shown in algorithm, first find the active parent node in the routing tree, and then assign channels to it. According to the delay requirements of information intelligent data collection, the node data packets that need to be forwarded are divided into three levels. According to different data priorities, a data transmission algorithm is proposed. Combining the above two algorithms, the channel allocation and data transmission based on data priority are finally achieved to improve the data transmission efficiency of the sensing nodes in the intelligent information collection.

In algorithm, parents represent the parent node set, interfp represents the interfered parent node, sm represents the STN ratio matrix, and the threshold refers to the interference signal-to-noise ratio threshold. As shown in Fig. 34.2, after initializing the parent node, it is necessary to traverse all the parent nodes, compare the signal-to-noise ratio thresholds one by one, and then traverse the nodes for classification. On the setting of the new parent node in the flowchart shown in Fig. 34.3, according to the example scheduling shown in Fig. 34.3, assuming that communication node

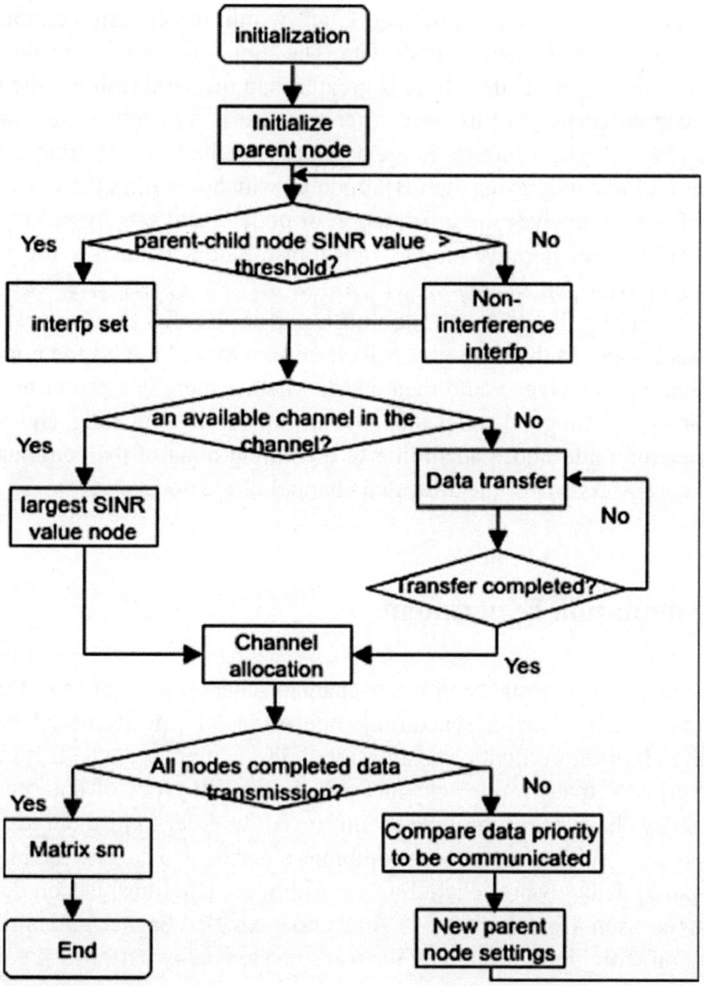


**Fig. 34.2** Flow chart of channel scheduling

**Fig. 34.3** Channel scheduling

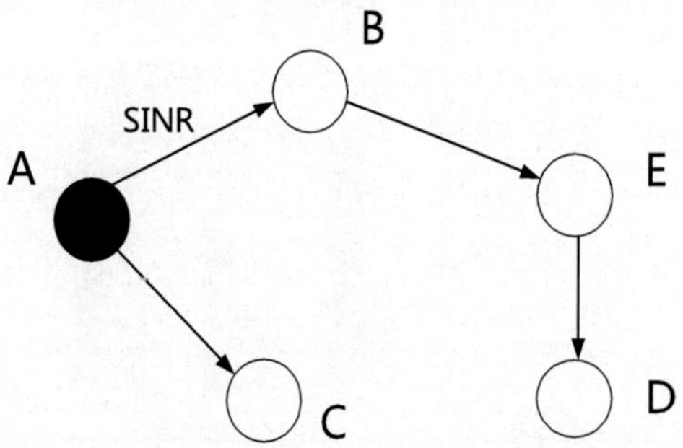

A is the parent node, and nodes B and C are within the communication radius of node A, it is calculated when A sends data. The signal-to-noise ratio values between A, B, and C are output. If the SINR is greater than the set threshold, the two nodes are allocated channels. At this time, after the data is transmitted according to the allocated channel, according to the data priority method in algorithm 2 (assuming that C data priority is greater than B), node E, which is within the communication radius of node C, receives the information of node C and sets its parent. The node position is C, and the iterative loop assigns transmission channels to the parent node in turn.

In this paper, the allocation of time slots is performed after the channel is allocated to the parent node. In the time gap, it is necessary to find out all the nodes that are communicating in advance, and then check whether there is a parent node for data communication among the nodes that are currently communicating, and find out the communication child nodes according to the parent node of the communication to realize the transmission of the allocated channel of the node.

## 34.4 Simulation Experiment

Delay perception performance of multi-channel scheduling algorithm: Demonstrate the effect of multi-channel scheduling, the effect of priority-based delay-aware scheduling algorithm on data transmission delay is mainly studied. It can be seen from experiment that in multi-channel scheduling, all types of data packet transmission delays have been significantly improved. In order to further reduce the data transmission delay, this experiment combines algorithm to discuss the impact of the priority-based delay-aware scheduling algorithm on data transmission delay.

It can be seen from Fig. 34.4b that under single-channel and multi-channel transmission with high traffic load, the transmission delay performance with delay-aware scheduling is significantly better than that without delay-aware scheduling. Comparing Fig. 34.4a, b, it can be seen that compared to the MHST algorithm, the

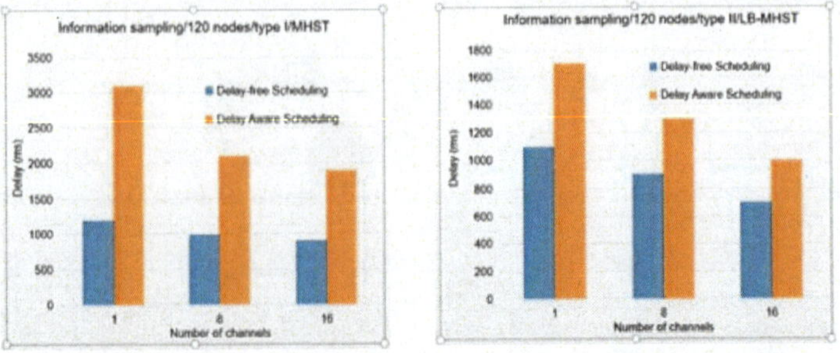


(a) Low load latency awareness (b) High load delay perception

**Fig. 34.4** Performance comparison of delay-aware scheduling under different loads (LB-MHST)

LB-MHST algorithm improves the delay performance by 12% in single-channel scheduling, the delay performance increases by 15.2% at 8 channels, and at 16 channels. The delay performance is improved by 18%.

## 34.5 Conclusion

Field measurement shows that low-power wireless communication in the application of infrastructure for intelligent information and intelligent collection will suffer from noise, interference, poor connectivity and fading due to the harsh and complex power grid environment. In addition, the influence of the proposed algorithm on the real-time communication of the grid master station is also analyzed. A large number of simulation experiments shows that the proposed algorithm can effectively reduce the communication delay in the smart grid environment. In future, we will make an in-depth investigation on the multi-channel scheduling of WSN in the worse smart grid environment. And study how to combine the scheduling strategy proposed in this paper with the weighted fair scheduling scheme to meet the fairness requirements in different intelligent information intelligent collection application scenarios.

**Acknowledgements** This work was financially supported by the science and technology project to State Grid Corporation "Research on the key technologies of power grid disaster intelligent perception and emergency command (5700-202019185A-0-0-00)".

## References

1. Liu, Z.Y.: Technology of Smart Grid, pp. 56–68. China Electric Power Press, Beijing (2010)
2. Wang, Y.M.: Research framework of robust smart grid technology standard system. Autom. Electr. Syst. **34**, 1–6 (2010)
3. Li, W., Ding, J., Yao, J.G.: Discussion in development of power system automation. Autom. Electr. Syst. **34**, 24–27 (2010)
4. Yaghmaee, M.H., Yousefi, Z., Zabihi, M. et al.: Quality of service guarantee in smart grid infrastructure communication using traffic classification. In: International Conference and Exhibition on Electricity Distribution, pp. 1–4. IET Press, Stockholm, Sweden (2013)
5. Xu, B.Y., Li, T.Y., Xue, Y.D.: Intelligent distribution network and distribution automation. Autom. Electr. Syst. **33**, 38–41 (2009)
6. Bari, A., Jiang, J., Saad, W. et al.: Challenges in the smart grid applications: an overview. Int. J. Distrib. Sensor Netw. **2014**, 1–11 (2014)
7. Musham, R., Kala, S.M., Muthyap, P. et al.: Near optimal channel assignment for interference mitigation in wireless mesh networks. In: 2016 IEEE International Conference on Advanced Networks and Telecommunications Systems, pp. 1–6. ANTS Press, Bangalore, India (2016)
8. Suryawanshi, R.: H-WSN with maximized QoS using secure data aggregation. In:2017 2nd International Conference on Contemporary Computing and Informatics, pp. 149–154. IC31 Press, Noida, India (2017)

9. Jayalakshmi, G., Nitya, N., Bsanakar, R.M.: Congestion based routing metric and channel assignment scheme for WMN. In: Proceeding of Second International Conference on Electrical, Computer and Communication Technologies, pp. 1–6. ICECCT Press, Tamil Nadu, India (2017)
10. Ahmed, A.A., Fisal, N.: A real-time routing protocol with load distribution in wireless sensor networks. Comput. Commun. **31**, 3190–3203 (2008)
11. Incel, O.D., Ghosh, A., Krishnamachari, B. et al.: Fast data collection in tree-based wireless sensor networks. IEEE Trans. Mob. Comput. **11**, 86–99 (2011)
12. Han, B., Han, T., Chen, L.J. et al.: Design and implementation of network model checking method for master station of power distribution network. Autom. Electr. Power Syst. **40**, 75–79 (2016)

# Chapter 35
# Anti-interference Performance of Transmit Diversity Techniques for Satellite System

**Ya Wen and Xin Wu**

**Abstract** In order to investigate the anti-interference performance of SORTD (space orthogonal-resource transmit diversity) for PUCCH (Physical Uplink Control Channel) format 3 in satellite system, this paper gives overhead analysis of transmit diversity for PUCCH format 3. The result of simulation shows that there is no strong requirement for enhancement of transmit diversity for PUCCH format 3 from the overhead point of view. Thus, a new scheme of transmit diversity for PUCCH format 3 should not be introduced in satellite system unless it brings little complexity on specification and implementation.

## 35.1 Introduction

In 3GPP Release 10, SORTD is supported as the transmit diversity scheme for PUCCH format 3 in satellite system. The amount of required PUCCH format 3 resources for SORTD is doubled compared to single antenna port transmission, and thus there was some concern on the PUCCH format 3 overhead with SORTD. One of the objectives in carrier aggregation work items is investigating the possible improvements including the anti-interference performance of transmit diversity scheme for PUCCH format 3 in satellite system [1].

In the RAN1 #66bis meeting, anti-interference performance of transmit diversity for PUCCH format 3 in satellite system was discussed. It was concluded that additional analysis on the overhead needed to be performed to identify whether enhancement of transmit diversity for PUCCH format 3 in satellite system is needed [2]. The contribution firstly discusses the method of performing the overhead analysis and then provides the overhead analysis of the anti-interference performance for PUCCH format 3 in satellite system.

---

Y. Wen (✉) · X. Wu
Chongqing College of Electronic Engineering (CQCEE), Chongqing 401331, China



L. C. Jain et al. (eds.), *Smart Communications, Intelligent Algorithms and Interactive Methods*, Smart Innovation, Systems and Technologies 257,
https://doi.org/10.1007/978-981-16-5164-9_35

The rest of this paper is organized as follows. In Sect. 35.2, we discuss the method of performing the overhead analysis of PUCCH format 3 in satellite system. Then the overhead analysis of transmit diversity for PUCCH format 3 in satellite system is described in detail in Sect. 35.3. Simulation results are presented and discussed in Sect. 35.4. This paper is concluded in Sect. 35.5 at last.

## 35.2 Discussion on Overhead Analysis Method

For the method of performing overhead analysis, the first question to be resolved is whether the PUCCH format 3 in satellite system overhead indicates the resources reserved semi-statically or the resources used in a given subframe. If all the unused PUCCH format 3 in satellite system resources can be used for other purposes like PUSCH (Physical Uplink Shared Channel) transmission in a given subframe, anti-interference performance should be evaluated based on the PUCCH format 3 in satellite system resources used in a given subframe [3]. However, usually, it is difficult to guarantee that all the unused PUCCH format 3 in satellite system resources can be used for PUSCH transmission because it would restrict the scheduling a lot. Thus, it is reasonable to evaluate the overhead based on the PUCCH format 3 in satellite system resources reserved semi-statically.

In Rel-10, the typical application scenarios of PUCCH format 3 in satellite system mainly include the cases with more than two configured serving cells. In time division dual (TDD), one or two configured serving cells with heavy DL configurations can also be the case. Whether PUCCH format 3 in satellite system will be configured to a user equipment (UE) mainly depends on the traffic requirement and the PUCCH channel condition. For traffic requirement, it is difficult to evaluate how frequent the typical application scenarios will happen because it depends on the traffic model. In addition, even though the typical application scenarios need to be configured because of the traffic requirement, the scenario or PUCCH format 3 in satellite system cannot be configured unless the performance target can be met under the PUCCH channel condition. Thus, for simplicity, we can evaluate the anti-interference performance of PUCCH format 3 in satellite system only from the PUCCH channel condition point of view. From this point of view, the following two aspects need to be evaluated first:

1. PUCCH geometry
   The PUCCH geometry determines the number of UEs that can be configured with PUCCH format 3 in satellite system. A system level simulation should be used to get the PUCCH SINR (signal to interference plus noise ratio) geometry. When performing the simulation, we should consider interference and UL PUCCH power control.
2. Link-level performance of PUCCH format 3 in satellite system
   Based on the PUCCH geometry and the link-level performance of PUCCH format 3 in satellite system with single antenna port (SAP) transmission and SORTD, we can get the percentage of UEs that can be configured with SAP

and the percentage of UEs that can be configured with SORTD, then we can get the PUCCH format 3 in satellite system resources that need to be reserved semi-statically for SORTD, assuming all the UEs with high enough geometry is configured with PUCCH format 3 in satellite system. Based on the PUCCH format 3 in satellite system resources reserved semi-statically for SORTD [4], we can analyze whether enhancement of transmit diversity for PUCCH format 3 in satellite system is needed or not.

## 35.3 Overhead Analysis of Transmit Diversity for PUCCH Format 3 in Satellite System

In this section, we evaluate the overhead of PUCCH format 3 in satellite system based on the method discussed in Sect. 35.2. PUCCH geometry and link-level performance of PUCCH format 3 in satellite system are evaluated firstly. Interference and PUCCH power control in our simulation are set as follows:

1. For interference
   The cases with maximum 1–5 UEs multiplexed on the same physical resource block (PRB) are simulated. In ideal condition, the inter-cell interference mainly comes from UEs using the same orthogonal cover code (OCC) as the desired UE. However, UEs using different OCCs also can be interferers considering some factors like the uplink (UL) reception window. In addition, the interference randomization for PUCCH format 3 in satellite system also has impact on the actual interference. For simplicity, we can evaluate the cases with different number of multiplexed UEs on the same PRB. The multiplexed UEs on the same PRB in other cells are the interferers. More multiplexed UEs can mean heavier interference to the desired UE.
2. For PUCCH power control

The setting of the UE transmission power $P_{\text{PUCCH}}$ in sub-frame $i$ is as follows:

$$P_{\text{PUCCH}}(i) = \min\{P_{\text{CMAX}} P_{0\_\text{PUCCH}} + PL + h(n_{\text{CQI}}, n_{\text{HARQ}}, n_{\text{SR}}) \\ + \Delta_{F\_\text{PUCCH}}(F) + \Delta_{\text{TxD}}\left(F'\right) + g(i)\} \quad (35.1)$$

where $P_{\text{CMAX}}$ is maximum transmitting power of UEs, $P_{0\_\text{PUCCH}}$ is Power reference value, $PL$ is the biggest loss of down channel measured, $h(n_{\text{CQI}}, n_{\text{HARQ}}, n_{\text{SR}})$ is the transmitting power offset of channel quality indicator (CQI), hybrid automatic repeat request (HARQ) and scheduling request (SR), $\Delta_{\text{TXD}}$ is the power offset of modulation encoding mode and data type, $g(i)$ is the adjustment value of closed-loop power control.

Then

$$P^1_{0\_PUCCH} = P_{0\_PUCCH} + h\left(n_{CQI}, n_{HARQ}, n_{SR}\right) + \Delta_{F\_PUCCH}(F) + \Delta_{TxD}\left(F'\right) + g(i) \tag{35.2}$$

And $P^1_{0\_PUCCH}$ is determined according to the following steps:

Get the target PUCCH received SINR (T-SINR). In our simulation, the target PUCCH received SINR is 3 dB.

Search the best $P^1_{0\_PUCCH}$, with which the most UEs could fulfil the target PUCCH received SINR. That is, the best $P^1_{0\_PUCCH}$ could ensure more UEs have a reliable PUCCH condition.

## 35.4 Simulations

The simulation results lead to the following observation: More multiplexed UEs on the same PRB, which means the inference is heavier, lead to worse PUCCH SINR CDF.

Figure 35.1 shows the PUCCH SINR cumulative distribution function (CDF) for the cases with different numbers of multiplexed UEs on the same PRB.

The link-level performance of PUCCH format 3 in satellite system with SAP and SORTD was evaluated in several contributions [5, 6]. Based on the simulation results in [5, 6], we can get the temporary minimum requirements for some typical

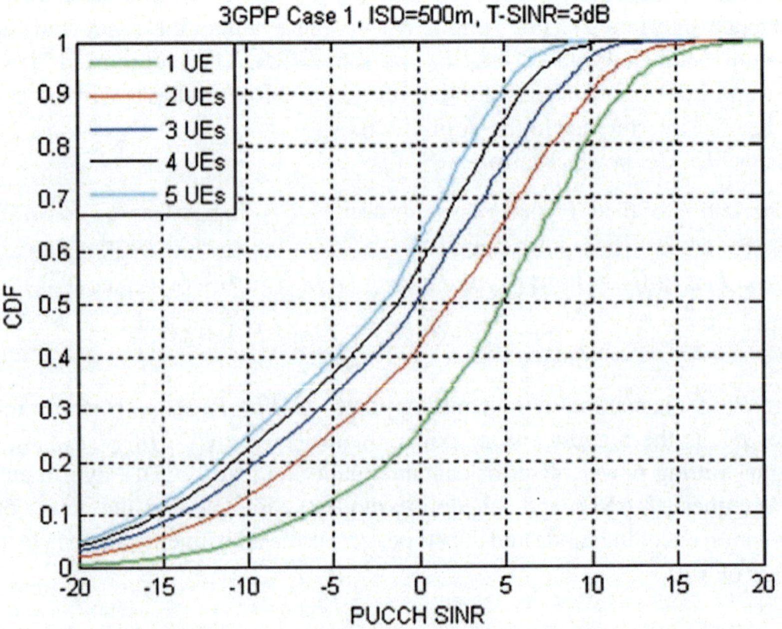


**Fig. 35.1** CDF for the cases with different number of multiplexed UEs on the same PRB

**Table 35.1** The minimum requirements (SNR [dB]) for different A/N payload sizes

| | 6 bits | 10 bits | 16 bits | 20 bits |
|---|---|---|---|---|
| Single antenna port (SAP) | –5.1 | –2.27 | 0.93 | 2.9 |
| SORTD | –6.91 | –5.1 | –2.39 | –1.39 |

**Table 35.2** Percentage of UE that can support PUCCH format 3 with SAP and SORTD

| | 6 bits (%) | 10 bits (%) | 16 bits (%) | 20 bits (%) |
|---|---|---|---|---|
| Percentage of UEs that can support PUCCH format 3 with SAP | 61–88 | 49–80 | 32–70 | 19–60 |
| Percentage of UEs that can support PUCCH format 3 with SORTD | 67–91 | 61–88 | 50–81 | 46–79 |

A/N payload sizes as shown in Table 35.1. Here the minimum requirement for each A/N payload size is the maximum required SINR among the simulation results in [5, 6].

Based on the simulation results for PUCCH SINR geometry and the anti-interference performance of PUCCH format 3 in satellite system, we can get the percentage of the UEs that support PUCCH format 3 in satellite system with SAP and the percentage of UEs that can support PUCCH format 3 in satellite system with SORTD as shown in Table 35.2.

If the anti-interference performance of PUCCH format 3 in satellite system with SAP can meet the target SINR requirement, it is reasonable to assume that SORTD will not be configured, because configuring SORTD requires doubled resource. Then we can further get the percentage of UEs that will be configured with PUCCH format 3 in satellite system with SAP and the percentage of UEs that will be configured with PUCCH format 3 in satellite system with SORTD as shown in Table 35.3.

According to Table 35.3, we can see that the percentage of the UEs that will be configured with PUCCH format 3 in satellite system with SORTD is small when the A/N payload size is 6 bits, thus the increase of the PUCCH format 3 in satellite system overhead brought by SORTD is marginal. When the A/N payload size is more than 6 bits, the percentage of the UEs that will be configured with SORTD is approximate 8% to 27%, which is large and thus it may have some impact on the PUCCH format 3 in satellite system overhead.

**Table 35.3** Percentage of UE that will be configured with PUCCH format 3 with SAP and SORTD

| | 6 bits (%) | 10 bits (%) | 16 bits (%) | 20 bits (%) |
|---|---|---|---|---|
| Percentage of UEs that will be configured with PUCCH format 3 with SAP | 61–88 | 49–80 | 32–70 | 19–60 |
| Percentage of UEs that will be configured with PUCCH format 3 with SORTD | 3–6 | 8–12 | 11–18 | 19–27 |

## 35.5 Conclusions

In this contribution, we discuss the method of performing the anti-interference performance of PUCCH format 3 in satellite system firstly and then provide the overhead analysis of transmit diversity for PUCCH format 3 in satellite system. Based on the evaluations and analysis, we can see that there is no strong requirement for enhancement of transmit diversity for PUCCH format 3 in satellite system from the overhead point of view. Thus, a new scheme of transmit diversity for PUCCH format 3 in satellite system should not be introduced unless it brings little complexity on specification and implementation.

**Acknowledgements** This work is supported by Research and Application of High Integrated Modular Avionic Systems (cstc2019jscx-zdztzxX0045).

## References

1. RP-110451: LTE Carrier Aggregation Enhancements, Nokia Corporation, Nokia Siemens Networks, Kansas City, USA, March 15th–18th (2011)
2. Iordache, A.V., Marghescu, I.: Transmit diversity in LTE network. Telecommunications Forum (TELFOR), 2013 21st. IEEE, pp. 145–148 (2013)
3. 3GPP.: Draft Report of TSG RAN WG1 #66bis v0.1.0, Zhuhai, China, 10th–14th October (2011)
4. 3GPP TS 36.213. Physical layer procedures
5. R1-113212: Transmit Diversity for PUCCH format 3, Huawei, HiSilicon, Zhuhai, China, October 10–14 (2011)
6. R1-113553: Transmit diversity for PUCCH Format 3, CATT, Zhuhai, China, October 10–14 (2011)

# Chapter 36
# Wireless Sensor Energy Harvesting and Management

**Huanan Zhang and Feng Wang**

**Abstract** This paper analyzes the energy harvesting and management of wireless sensor network nodes and studies the principle, technology and method of energy collection of wireless sensor network nodes. Energy management of wireless sensor network nodes should be solved from two aspects of energy saving and energy supply. The harvesting principles and methods of various energy sources in the environment, including solar energy, wind energy, sound energy, vibration energy, thermoelectricity and electromagnetic field energy, are analyzed. Sensor nodes should absorb energy from the environment in as many ways as possible to ensure long-term, stable and reliable operation of sensor nodes.

## 36.1 Introduction

Wireless sensor network (WSN) is self-organizing, miniaturized and aware of the outside world. It integrates many new technologies, such as sensors, embedded systems, communication and power supply. Wireless sensor network nodes are usually composed of sensors and communication circuits and can be placed in factories, farmland or battlefield and other dangerous or inconvenient places to complete positioning, measurement, control and other functions. Can at any time, any place through data collection, processing, analysis, dissemination to provide a new information channel, so that people get more detailed and reliable information, in industrial control, traffic management, environmental monitoring, space exploration and other fields has a potential and extensive application needs [1].

Wireless sensor network node is stationary and, possibly in the wild environment, is not allowed to change the battery; therefore, wireless sensor network node is the energy management problem for prolonging the lifetime of wireless sensor

H. Zhang
School of Data and Computer Science, Guangdong Peizheng College, Guangzhou 510830, China

F. Wang (✉)
School of Computer Engineering, Hubei University of Arts and Science, XiangyangHubei Province 441053, China
e-mail: wangfeng@hbuas.edu.cn

 
L. C. Jain et al. (eds.), *Smart Communications, Intelligent Algorithms and Interactive Methods*, Smart Innovation, Systems and Technologies 257,
https://doi.org/10.1007/978-981-16-5164-9_36

network application and reduce the cost of the key, to become one of the core problems in the research of wireless sensor network, involves two aspects, namely energy consumption and energy problems. Therefore, in order to solve the problem of energy management of wireless sensor network nodes, it is necessary to conduct in-depth and detailed research from these two aspects. At present, there are many researches on solving the problem of energy consumption. For example, in order to effectively utilize the existing energy resources and extend the life cycle of the network, various optimized routing communication protocols are studied. It is impossible for wireless sensor network nodes to work normally for a long time only by various optimization and consumption reduction methods. When the energy consumption has been reduced to a certain limit by various measures, no matter how hard people try, they will not get a better effect. Therefore, we must study from the perspective of energy supply and take effective methods to provide continuous energy supply for wireless network sensors. Wireless sensor network nodes can also obtain and store energy from their environment, so research on how to effectively collect and store energy from the environment and energy collection methods have attracted more and more researchers' attention [2].

## 36.2 Requirements for Energy Harvesting

Wireless sensor networks generally consist of a large number of sensor nodes scattered over a certain area, usually powered by batteries. However, due to the limitations of node size, the configured battery can provide very limited energy. At the same time, because the sensor nodes are often in harsh environments or inaccessible environments, and the number of sensor nodes is very large, it is not possible to replace the battery for each node. In order to prolong the life of sensor nodes, it is necessary to take various methods to replenish the energy of sensor nodes. The sensor nodes harvest energy from the environment, known as energy harvesting. Wireless sensor nodes collect all available energy from the environment and store it. When the node needs energy, the energy is taken out of the storage unit and transformed to get the energy needed on the node [3].

According to the environment of sensor nodes, the energy collected in the environment is also different, so the energy collection method of a single energy is difficult to ensure that all nodes in the wireless sensor network can reliably obtain the required energy. Therefore, it is necessary to set up two or more energy collection methods for each sensor node, which requires that a comprehensive energy collection power supply should be configured as far as possible in a wireless sensor network node with limited space according to the possible energy types in the working environment of the node [4].

## 36.3 Analysis of Energy Harvesting Technology

There may be a variety of potential energy available in the physical space we live in, such as solar (light) energy, wind energy, heat energy, mechanical vibration energy, sound energy, electromagnetic field energy and so on. How to collect and store the energy in small sensor network nodes is one of the focus problems that many scientists have studied hard in recent years, and some progress has been made at present. Among them, the energy harvesting technology using mechanical vibration and light energy is more studied, and the products of related devices appear, which has a good application prospect.

### *36.3.1 Solar (Light) Energy Collection*

New advances in optoelectronic materials make light energy collection, a new method of power source for wireless network sensors. The installation and operation cost of optoelectronic components can also be greatly reduced with large-scale applications. The basic principle of photoelectric acquisition is to use photoelectric materials to absorb a large number of photons. If enough photons can activate the electrons in the photocell, electrons can be obtained after proper structural design. Photoelectric components act as decoders, generating voltages under light irradiation, which, combined with the corresponding adjustment and storage circuits, can supply power to the load. The amount of electric quantity is a function of the collected light energy. To obtain more electric quantity, the photoelectric element is usually placed in a well-lit environment and the area of illumination is increased. Usually photovoltaic cells can produce voltage DC 0.5 V, but the actual voltage output varies with the operating temperature. Generally speaking, the lower the temperature, the higher the output voltage, the stronger the light, the greater the current output. Optoelectronic technology has evolved from the initial manufacture of silicon crystals to today's deposition of tiny particles on photosensitive substrates. This new material can be used indoors or outdoors; it is easy to install in weight, and is less affected by ambient temperature, making it ideal for powering small, remote sensors. Solar collection power sources using super capacitors as energy buffers are a universal and representative model that is well suited for a wide range of sensor network applications, as shown in Fig. 36.1.

### *36.3.2 Wind Energy Harvesting*

The wind energy collection device consists of a rotating shaft and a number of blades. One end of the blade is fixed on the rotating axis and radially extends to all sides along the rotating axis. The blades form an oblique Angle to the shaft. When the wind

**Fig. 36.1** Energy harvesting sensor node

blows head-on against the blade, the blade rotates on its axis under the action of the wind. The shaft rotates, collecting energy from the wind. The wind energy collecting effect is related to the wind side. The effect of wind from different directions is very different. The wind force also has a big impact on wind harvesting.

### *36.3.3 Harvesting of Thermoelectric Energy*

Research on thermoelectric technology began in the 1940s and reached its peak in the 1960s with the successful realization of long-term power generation on spacecraft. The thermoelectric generator has the characteristics of small size, light weight, no vibration, no noise, reliable performance, less maintenance and can work for a long time in extreme harsh environment. It is suitable for small power supply of less than 5 W, for a variety of unsupervised sensors, small short-range communication devices and medical and physiological research instruments. At present, the relevant products have entered the practical stage. In recent years, thermoelectric generator has also shown a good application prospect in civil use. Through the in-depth study of thermoelectric conversion materials and the development of new materials, it has become the core content of thermoelectric technology research to continuously improve the thermoelectric performance and improve the electric output power under the condition of constant heat source. Scientists have invented a new type of battery that uses human temperature differences to generate electricity, which could provide long-term power for portable miniature electronic instruments, eliminating the trouble of charging or replacing batteries. As long as there is a temperature difference of 5 °C between the human skin and clothes, you can use this battery to provide enough energy for an ordinary watch.

### 36.3.4 Collection of Sound Energy

Sound energy is a form of energy. Its essence is the transfer and transformation of mechanical energy through the media and in the form of wave after the object vibrates. Conversely, the transfer and transformation of other energies can also be reduced to mechanical energy to produce sound. The change can be reversed. To convert sound energy into electricity when sound waves encounter obstacles, it is possible to design and build resonators that collect sound when the input noise energy they collect is converted into electricity storage.

Conventional engines are limited by thermodynamics and engine complexity. The most efficient engines are typically giant turbines used in power stations. Small sound-energy engines are up to 10% more powerful than the largest, most efficient turbines and have no moving parts and no maintenance.

### 36.3.5 Magnetic Energy Harvesting

In the magnetostatic case, electric current and magnetic field always exist together. Thus,Therefore, the magnetic energy can be regarded as related to electric currents and stored in the magnetic field. However, scientific practice has proved that magnetic field is a special form of matter, which can exist without electric current. The changing electric field can also produce a magnetic field, which also has energy and its field energy density is the same as that of magnetostatic. In general, changing electromagnetic fields propagate in the form of waves, accompanied by energy transfer.

The electromagnetic energy recovery system is mainly composed of receiving antenna, rectifier circuit and energy management module, among which the performance of antenna and rectifier circuit is the key factor affecting the performance of the whole system. In this system, the receiving antenna determines how much electromagnetic energy can be collected, and the rectifying circuit determines how much direct current energy can be converted into electromagnetic energy that can be directly used. On the one hand, the electromagnetic energy distributed in the environment has location, frequency uncertainty and power inhomogeneity, and the current design of electromagnetic energy recovery system mainly uses single-frequency or multi-frequency high-gain directional antenna to solve the recovery problem. However, this scheme often requires high alignment of the receiving and receiving antennas, and the actual relative position between the energy source and the receiving antenna is often uncertain. On the other hand, there are two main problems in the existing rectifier circuit schemes: impedance matching, secondary rectification and impedance compression are adopted to improve the conversion efficiency, but the input power is too high when the optimal rectification efficiency is achieved; The nonlinear action of diodes makes the input power range of the rectifier circuit narrow when efficient rectification. Therefore, for the practical application scenarios with uncertain electromagnetic energy, the existing recovery schemes still fail to solve the current problems well [5].

### *36.3.6 Comprehensive Harvesting of Energy from Various Sources*

We live in an environment in which there are many forms of energy. Energy takes different forms in different environments. In order for each sensor node to obtain as much energy as possible from its environment, it is necessary to design such an energy collection system, which cannot collect energy from only one kind of energy, otherwise, the node will not work reliably for a long time once the environment is short of this kind of energy. Therefore, it is necessary to integrate multiple energy harvesting methods on each node. Of course, the difficulties are not difficult to imagine, mainly manifested in the following aspects: (1) various energy recovery technologies and methods are not mature at present, and researchers need to carry out a lot of innovative research; (2) Various energy harvesting components must meet the stringent size requirements of sensor network nodes; (3) Ensure that the various energy harvesting components work in concert and that the collected energy is effectively stored.

## 36.4 Energy Management

The energy collected must be managed effectively so that it can be continuously and effectively supplied to the sensor, otherwise the energy collected will be useless. Power management is a hot topic in energy utilization. Dynamic power management is a potential method to save power. The storage element of power supply is battery or capacitor. Chemical battery has become the first choice in network sensor because of its stable and effective storage. The limitation of collecting energy into the battery is the battery life problem. How to make the battery life meet the user's acceptable expectation and adopt the correct charging method to extend the battery life is the key problem. Dynamic power management and dynamic voltage dispatching are adopted in power management of networked sensors, which is an effective design method to reduce the energy consumption of the system. Dynamic power management (DPM) is a design method to reduce power consumption by dynamically allocating system resources and completing system tasks with the least components or the least workload.

Dynamic power management technology includes a series of methods to enable the system to achieve efficient energy saving. These methods control the "power management" of whether system components enter a low-power state when they are idle. The basic premise of the application of dynamic power management technology is that the system components have different workload during the working time. This is true of most systems. Another premise is that fluctuations in the workload of systems and components can be predicted with some degree of confidence. In this way, it is possible to switch between states of energy consumption, and the system cannot consume excessive energy within the observed and predicted time of the workload. In the network sensor power management, its working characteristics are very suitable for dynamic power management. Because the nodes of the sensor network are not working all the time, they spend a lot of time in sleep mode, which requires low power consumption. When the sensor collects the data, the controller processes the data and carries on the data communication, the system power consumption increases. When entering idle, the device can be shut down and enter the low-energy sleep state; when the request is received again, the device is evoked [6].

In a dynamic power management system, the working states of different components must dynamically adapt to different performance requirements. Only in this way can the energy wasted in idle time or in useless components be minimized. For the judgment of power management time, it is necessary to use a variety of prediction methods, such as static prediction method and dynamic prediction method. It is necessary to predict the coming workload according to the historical workload and decide whether to change the working state. The main principle of dynamic voltage scheduling is to adjust the supply voltage dynamically based on the load state to reduce the power consumption of the system, which has been applied in portable personal mobile devices. Sensor nodes on the embedded operating system is responsible for scheduling request to receive services from different task queue, and real-time monitor the processor utilization and task queue length, the load observer based on the sequence of these two parameters value computation load of nominal value, DC/DC converter with reference to the value of the output amplitude of voltage, support the work of the processor. This actually constitutes a closed-loop feedback system, and the method of control theory can be used to design each module.

## 36.5 Conclusion

Wireless sensor network is widely used. In order to prolong the life cycle of sensor nodes, improving the energy efficiency of sensor nodes has always been an important research topic. In this paper, the energy collection and management of wireless sensor network nodes are analyzed and discussed. Solar energy, wind energy, sound energy, magnetic energy and so on are introduced and analyzed. In order to make the sensor nodes work stably and reliably for a long time, the most effective method is to adopt various methods to obtain energy from the environment and reduce energy consumption and improve energy efficiency through energy management.

## References

1. Wang, Q., Dai, H. N., Li, X., Wang, H., Xiao, H.: On modeling eavesdropping attacks in underwater acoustic sensor networks. Sensors **16**(5), 721 (2016)
2. Wang, Z., Han, G., Qin, H, Zhang, S., Sui, Y.: An energy-aware and void-avoidable routing protocol for underwater sensor networks. IEEE Access, **8**(6), 7792–7801 (2018)
3. Gherbi, C., Aliouat, Z., Benmohammed, M.: An adaptive clustering approach to dynamic load balancing and energy efficiency in wireless sensor networks. Energy, **114**, 647–662 (2016)
4. Long, J., Dong, M. X., Ot, K., Liu, A. F.: Green TDMA scheduling algorithm for prolonging lifetime in wireless sensor networks. IEEE Syst. J. 1–10 (2015)
5. Mohanty, P., Kabat, M.R.: Energy efficient reliable multi-path data transmission in WSN for healthcare application. Int. J. Wirel. Inf. Netw. **23**, 162–172 (2016)
6. Lal, C., Petroccia, R., Pelekanakis, K., Conti, M., Alves, J.: Toward the development of secure underwater acoustic networks. IEEE J. Ocean. Eng. **42**, 1075–1087 (2017)

# Chapter 37
# Stacked Sparse Autoencoder-Based WLAN Fingerprinting Localization

**Junhang Bai, Yongliang Sun, and Quan Zhang**

**Abstract** Currently, deep learning algorithms have been used in a variety of fields because their performance usually outperforms traditional machine learning algorithms. In this study, we present a stacked sparse autoencoder (SSAE)-based wireless local area network (WLAN) fingerprinting localization algorithm. The hidden layer dimension of a basic autoencoder (AE) is limited by its input dimension, which could affect its performance. By contrast, the hidden layer dimension of a sparse autoencoder can be set to be greater than its input dimension, so we train two sparse autoencoders in turn for unsupervised received signal strength indicator (RSSI) feature learning. Then we stack the encoders and also add a linear regression layer as an output layer to the SSAE structure. We obtain the SSAE fingerprinting localization algorithm after supervised fine-tuning. We evaluate our presented SSAE localization algorithm in an actual indoor office scenario. The localization results show that our presented SSAE fingerprinting algorithm achieves a better performance compared with some other popular fingerprinting localization algorithms.

## 37.1 Introduction

With the development of mobile communication and proliferation of smartphone applications, location-based service (LBS) has been paid intensive attentions and shown great application potentials in the venues like shopping mall, hospital, and so on [1]. Due to the limited performance of satellite and cellular-based localization systems indoors, different localization systems for indoor scenarios have been implemented using ultrasound, Bluetooth, ultra-wideband (UWB), and wireless local area network (WLAN) [2]. Among these existing systems, WLAN fingerprinting localization has been favored because of the widely available WLAN access points (APs) and satisfactory localization performance [3]. At present, various fingerprinting localization algorithms have been utilized. For instance, weighted k-nearest neighbors (WKNN) and multi-layer perceptron (MLP) [4]. Nowadays, deep learning algorithms

J. Bai · Y. Sun (✉) · Q. Zhang
School of Computer Science and Technology, Nanjing Tech University, Nanjing 211816, China
e-mail: syl_peter@163.com



L. C. Jain et al. (eds.), *Smart Communications, Intelligent Algorithms and Interactive Methods*, Smart Innovation, Systems and Technologies 257,
https://doi.org/10.1007/978-981-16-5164-9_37

have demonstrated its strong power in massive data processing, feature extraction, and so on [5] and has been applied in the field of WLAN fingerprinting localization.

[6, 7]. Thus, in this study, we present a stacked sparse autoencoder (SSAE)-based WLAN fingerprinting localization algorithm.

## 37.2 System Model

### *37.2.1 System Overview*

Just like the basic fingerprinting localization system, the SSAE-based WLAN fingerprinting localization system also can be divided into two steps that are the offline and online step. Firstly, in the offline step, reference points (RPs) are selected and their locations are recorded. At these RPs, received signal strength indicator (RSSI) are recorded to establish a fingerprint database called radio-map. Specifically, the SSAE fingerprinting localization algorithm is trained by employing the radio-map. The SSAE structure can be constructed by stacking the encoders of sparse autoencoders after the unsupervised feature learning, and then its weights and biases are optimized through the supervised fine-tuning. In the online step, the measured RSSI vector can be inputted into the trained SSAE to calculate the user's location vector.

### *37.2.2 Fingerprinting Localization Using SSAE*

Unsupervised Feature Learning. A basic AE is a kind of neural network that has one hidden layer. It consists of an encoder $\mathrm{E}(r)$ and a decoder $\mathrm{D}(r)$. Through training, the decoder output vector is expected to approach to the encoder input vector, that is $\mathrm{D}[\mathrm{E}(r)] \approx r$. The encoder output vector $\boldsymbol{a}$ and the decoder output vector $\hat{\boldsymbol{r}}$ are computed as follows:

$$\begin{cases} \boldsymbol{a} = \mathrm{E}(\mathbf{w}\boldsymbol{r} + \boldsymbol{b}) \\ \hat{\boldsymbol{r}} = \mathrm{D}\left(\mathbf{w}^{\mathrm{T}}\boldsymbol{a} + \boldsymbol{b}'\right) \end{cases} \tag{37.1}$$

where $\boldsymbol{r}$ is the input vector of the AE, $\mathbf{w}$ is the weight matrix between the input layer and the hidden layer, $\mathbf{w}^{\mathrm{T}}$ is the transposition of $\mathbf{w}$ and is the weight matrix between the hidden layer and the output layer, $\boldsymbol{b}$ is the bias vector of the hidden layer, $\boldsymbol{b}'$ is the bias vector of the output layer.

We employ the back propagation (BP) to update weight matrices and bias vectors of the AE to minimize the loss function $J_{\mathrm{AE}}$ between the $\boldsymbol{r}$ and $\hat{\boldsymbol{r}}$. The optimized weight matrices and bias vectors of the AE are obtained when the loss function $J_{\mathrm{AE}}$ is reduced. The $J_{\mathrm{AE}}$ can be computed as follows:

$$J_{\mathrm{AE}} = \frac{1}{N}\sum_{n=1}^{N} \left\| \boldsymbol{r}_n - \hat{\boldsymbol{r}}_n \right\|_2 \tag{37.2}$$

where $\boldsymbol{r}_n$ is the $n$th input vector, $\hat{\boldsymbol{r}}_n$ is the $n$th output vector, $N$ is the number of input vectors, and $\|\cdot\|_2$ is 2-norm.

An AE can be used for dimension reduction or feature extraction of RSSI data. In this progress, the dimension reduction of the hidden layer makes the network learn an RSSI compressed representation. If only a few APs can be detected by a terminal device at the same time, the dimension of input RSSI vector could be limited. Then the dimension reduction of the hidden layer of the AE will have a serious influence on the network performance. To solve this problem, we exploit a sparse autoencoder for unsupervised feature learning. The dimension of the hidden layer of the sparse autoencoder is not limited by the dimension of input RSSI vector, which will improve the performance of the supervised fine-tuning. A sparse autoencoder can be derived through adding a sparsity restriction to an AE's hidden layer. Its hidden layer dimension can be set to be greater than its input dimension.

The sparsity restriction can be considered that the hidden layer neurons are 0 in most cases. On this basis, we add a sparsity penalty factor $J_{\mathrm{KL}}$ for restricting the sparsity to the loss function $J_{\mathrm{AE}}$.

First, we compute the mean of the activation $\rho'_q$ of the $q$th neuron in the hidden layer of all the $N$ input vectors of a sparse autoencoder. It can be calculated as follows:

$$\rho'_q = \frac{1}{N}\sum_{n=1}^{N} h_{nq} \tag{37.3}$$

where $h_{nq}$ is the $q$th hidden layer neuron output of the $n$th RSSI vector.

Then we use a sparsity parameter $\rho$ closed to 0 to approximately enforce the sparsity restriction $\rho'_q = \rho$. Meanwhile, we also use relative entropy called Kullback–Leibler (KL) divergence as a regularization term, with which the sparsity penalty factor $J_{\mathrm{KL}}$ is computed as follows:

$$J_{\mathrm{KL}} = \sum_{q=1}^{Q} \left( \rho \log \frac{\rho}{\rho'_q} + (1-\rho) \log \frac{1-\rho}{1-\rho'_q} \right) \tag{37.4}$$

where $Q$ is the number of neurons in the hidden layer.

Besides, we also add a weight penalty factor $J_{\mathrm{WP}}$ to the loss function to avoid over-fitting and the weight penalty factor $J_{\mathrm{WP}}$ is calculated as follows:

$$J_{\mathrm{WP}} = \sum_{l=1}^{L}\sum_{q=1}^{Q} w_{lq}^2 \tag{37.5}$$

where $w_{lq}$ is the weight between the $l$th input and the $q$th hidden layer neuron, $L$ is the dimension of the input vector.

Thus, the sparse autoencoder's loss function $J_{\mathrm{LF}}$ can be calculated as follows:

$$J_{\mathrm{LF}} = J_{\mathrm{AE}} + k_1 J_{\mathrm{KL}} + k_2 J_{\mathrm{WP}} \tag{37.6}$$

where $k_1$ is the weighted parameter of the sparsity penalty factor and $k_2$ is the weighted parameter of the weight penalty factor.

An SSAE can contain multiple encoders of sparse autoencoders. In this study, we establish an SSAE with sparse autoencoder 1 and sparse autoencoder 2. We firstly train sparse autoencoder 1 using the standardized RSSI data. After that, we also train sparse autoencoder 2 using the hidden layer neuron outputs of the sparse autoencoder 1. We also add a linear regression layer as the output layer to the trained SSAE structure. Figure 37.1 reveals the construction of the presented SSAE.

Supervised Fine-Tuning. In order to fine-tune the SSAE, we exploit the RPs' RSSI and location vectors as the SSAE inputs and outputs, respectively. The output vector $\boldsymbol{y}$ of the SSAE is calculated as follows:

$$\boldsymbol{y} = \mathbf{w}^{(3)} \boldsymbol{a}^{(2)} + \boldsymbol{b}^{(3)} \tag{37.7}$$

where $\boldsymbol{a}^{(2)}$ is the vector in the hidden layer 2, $\mathbf{w}^{(3)}$ is the weight matrix between the vector in the hidden layer 2 and the output vector, $\boldsymbol{b}^{(3)}$ is the bias vector of the output layer.

We also employ the BP to fine-tune weight matrices and bias vectors of the SSAE to minimize the loss function $J_{\mathrm{SSAE}}$. The loss function $J_{\mathrm{SSAE}}$ is computed as follows:

$$J_{\mathrm{SSAE}} = \frac{1}{N} \sum_{n=1}^{N} \left\| \boldsymbol{L}_n - \boldsymbol{y}_n \right\|_2 \tag{37.8}$$

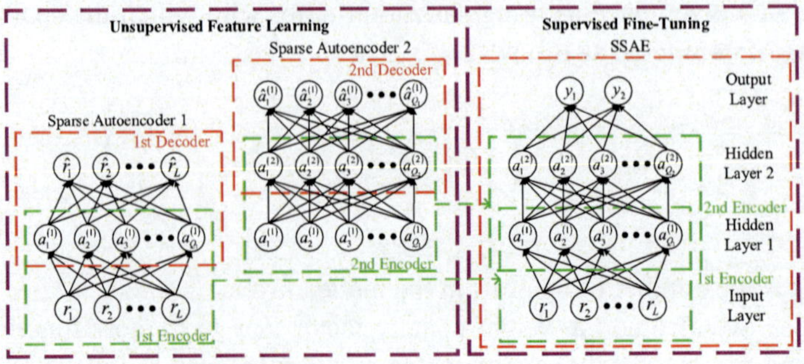


**Fig. 37.1** The construction of the presented SSAE

where $\boldsymbol{L}_n$ is the real location of $n$th RSSI vector, and $\boldsymbol{y}_n$ is the calculated location by our presented SSAE using $n$th RSSI vector.

## 37.3 Experimental Setup and Results

### *37.3.1 Experimental Setup*

We obtained the RSSI and location data in an actual office scenario with 20.4 m width × 51.6 m length. The plan is shown in Fig. 37.2. We mounted 7 TP link TL-WR845N APs at 2.2 m height. In order to establish the radio-map, we chose 116 RPs totally in the rooms and corridor, and then we recorded 120 RSSI vectors at every RP. Similarly, we also chose 90 test points (TPs) in the scenario and recorded 60 RSSI vectors at every TP to test our presented algorithm. We utilized a Meizu M2 smartphone at 1.2 m height for RSSI measurement with a rate of 1 RSSI vector per second.

In the unsupervised feature learning phase, we set up the sparsity parameter $\rho = 0.05$ of the two sparse autoencoders. We set up the weighted parameters $k_1 = 1$ and $k_2 = 0.001$. At the same time, we generated the weight matrices of all the sparse autoencoders using a Gaussian distribution with mean $\mu_1 = 0$ and variance $\sigma_1^2 = 0.01$, and set each component of the bias vectors of all the sparse autoencoders equal to 0.1. Additionally, we set up the neuron numbers $Q_1 = 100$ and $Q_2 = 200$ in the hidden layer 1 and 2, respectively. Also, we set up the iterative epochs $ie_1 = 300$ and learning rate $lr_1 = 0.01$.

In the supervised fine-tuning phase, we generated the output layer weight matrix of the SSAE using a Gaussian distribution with mean $\mu_2 = 0$ and variance $\sigma_2^2 = 1$, and set each component of the bias vectors of the output layer equal to 0.1. To fine-tune the weights and biases of the SSAE, we also set up the iterative epochs $ie_2 = 300$, dropout rate $dr = 0.2$, and learning rate $lr_2 = 0.005$.

**Fig. 37.2** The experimental scenario plan

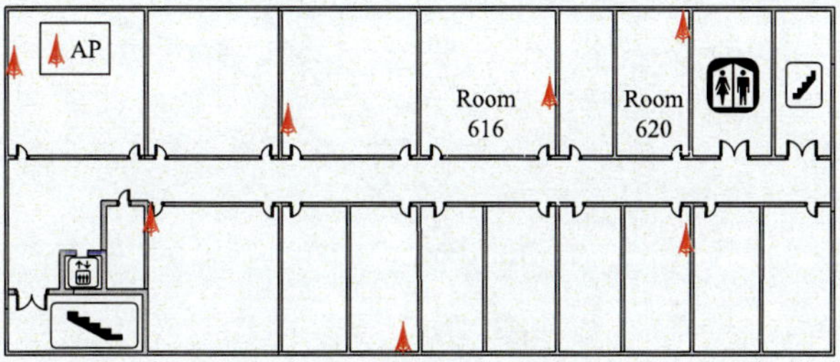

**Table 37.1** Performance comparison of different fingerprinting algorithms

| Algorithm | Mean error (m) | Cumulative probability (%) | |
|---|---|---|---|
| | | Within 2 m error | Within 3 m error |
| WKNN | 4.20 | 37.9 | 54.4 |
| MLP | 2.55 | 45.1 | 69.3 |
| DNN | 2.45 | 48.9 | 72.2 |
| SAE | 2.34 | 51.4 | 73.0 |
| SSAE | 2.23 | 56.7 | 75.1 |

## *37.3.2 Experimental Results*

We compare our presented SSAE fingerprinting algorithm with some other popular fingerprinting algorithms: WKNN, MLP, deep neural network (DNN), and stacked autoencoder (SAE). Regarding the WKNN, we select 7 nearest RPs for localization. We set up the hidden layer neuron number of the MLP $Q_{\mathrm{MLP}} = 100$. The DNN, SAE, and SSAE have the same parameters and network structure. A total of 5400 testing RSSI vectors are used for testing these algorithms, and the computed localization results are listed in Table 37.1. The mean testing errors of the WKNN, MLP, DNN, SAE, and presented SSAE are 4.20 m, 2.55 m, 2.45 m, 2.34 m, and 2.23 m, respectively. The presented SSAE has the minimum mean error. Meanwhile, the cumulative probabilities of different fingerprinting algorithms are shown in Fig. 37.3. The cumulative probabilities of the SSAE with a localization error of 2 m and 3 m can reach 56.7% and 75.1%, respectively. The SSAE outperforms the other fingerprinting algorithms.

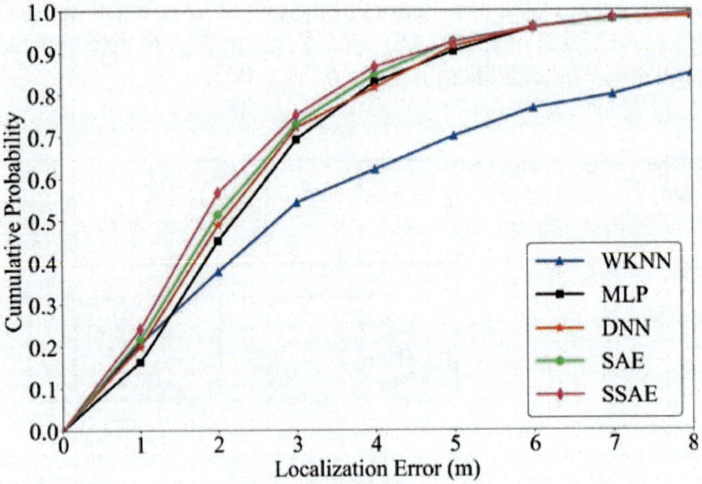


**Fig. 37.3** The cumulative probability curves of different fingerprinting algorithms

## 37.4 Conclusion

We present an SSAE-based WLAN fingerprinting localization algorithm. Because the hidden layer dimension of a sparse autoencoder is not limited by the dimension of input RSSI vector, we exploit the sparse autoencoder and train two sparse autoencoders for unsupervised feature learning. Then we stack the encoders of the two sparse autoencoders to construct an SSAE structure and add a linear regression layer to the SSAE structure for fingerprinting localization. We train the SSAE using BP algorithm for supervised fine-tuning. The localization results reveal that our presented SSAE fingerprinting algorithm has the minimum mean testing error of 2.23 m compared with some other popular fingerprinting algorithms.

**Acknowledgements** This work was supported by the National Natural Science Foundation of China under Grant 61701223 and also by the Natural Science Foundation of Jiangsu Province under Grant BK20171023.

## References

1. Guo, X., Ansari, N., Li, L., Duan, L.: A hybrid positioning system for location-based services: design and implementation. IEEE Commun. Mag. **58**(5), 90–96 (2020)
2. Gu, Y., Lo, A., Niemegeers, I.: A survey of indoor positioning systems for wireless personal networks. IEEE Commun. Surv. Tutor. **11**(1), 13–32 (2009)
3. Zafari, F., Gkelias, A., Leung, K.K.: A survey of indoor localization systems and technologies. IEEE Commun. Surv. Tutor. **21**(3), 2568–2599 (2019)
4. Sun, Y., Meng, W., Li, C., Zhao, N., Zhao, K., Zhang, N.: Human localization using multi-source heterogeneous data in indoor environments. IEEE Access **5**, 812–822 (2017)
5. Ma, X., Yao, T., Hu, M., Dong, Y., Liu, J.: A survey on deep learning empowered IoT applications. IEEE Access **7**, 181721–181732 (2019)
6. Khatab, Z.E., Hajihoseini, A., Ghorashi, S.A.: A fingerprint method for indoor localization using autoencoder based deep extreme learning machine. IEEE Sensors Letters **2**(1), 1–4 (2018)
7. Li, Y., Hu, X., Zhuang, Y., Gao, Z., Zhang, P., El-Sheimy, N.: Deep reinforcement learning (DRL): another perspective for unsupervised wireless localization. IEEE Internet Things J. **7**(7), 6279–6287 (2020)

# Chapter 38
# A Gridless Joint DOA and Range Estimation Method for FMCW MIMO Radar

**Jingyu Cong, Tao Lu, Zhang Bin, Xiaoli Jing, and Xianpeng Wang**

**Abstract** The traditional two-dimensional multiple signal classification (2D-MUSIC) has low accuracy and high computational complexity which is not effective in the joint direction-of-arrival (DOA) and range estimation with frequency-modulated continuous wave multiple-input multiple-output (FMCW MIMO) radar. In this paper, a gridless algorithm for joint DOA and range estimation is proposed. Firstly, the atomic norm denoising method is used to reconstruct the low-noise covariance of the received signal. Next, the DOA of targets is calculated by the total least square estimation of signal parameters via rotational invariance techniques (TLS-ESPRIT) algorithm and then reconstruct the steering vector matrix. Finally, the range of targets is obtained by the least square method. At the same time, the proposed method ensures the automatic matching of DOA and range. The effectiveness of the algorithm is proved by experimental data of 2.4G band FMCW MIMO radar which is more excellent than 2D-MUSIC.

## 38.1 Introduction

Frequency-modulated continuous wave (FMCW) radars has been successful in many applications [1]. FMCW multiple-input multiple-output (MIMO) radar has an equivalent virtual antenna array which is pairing transmitting and receiving elements as virtual element independently [2], which enlarges the array aperture and enhances the estimation performance.

As a solution to estimate DOA and range estimation jointly, two-dimensional (2D) algorithms have been proposed [2–4]. In [2], 2D fast Fourier transformation (FFT) which is a fast algorithm is used to estimate direction-of-arrival (DOA) and range, unfortunately, limited by the Rayleigh criterion of FMCW MIMO system,

J. Cong · Z. Bin · X. Jing · X. Wang (✉)
State Key Laboratory of Marine Resource Utilization in South China Sea and School of Information and Communication Engineering, Hainan University, Haikou 570228, China
e-mail: wxpeng1986@126.com

T. Lu
Beijing Institute of Aerospace Engineering, Beijing 100076, China

 
L. C. Jain et al. (eds.), *Smart Communications, Intelligent Algorithms and Interactive Methods*, Smart Innovation, Systems and Technologies 257,
https://doi.org/10.1007/978-981-16-5164-9_38

and the DOA resolution and range resolution are not very desirable. In [3], a 2D multiple signal classification (2D-MUSIC) method for the DOA and range imaging was presented, which was able to obtain a good performance in experiments, but just like the algorithm in [4], a huge covariance matrix and 2D spectral peak search are needed that lead to high computational complexity. In [4], a CAPON-based 2D algorithm has been proposed; however, due to the matrix inversion, computational complexity is $O(M^3)$, and this method is hard to meet the calculation speed requirements either.

Thus, this paper proposed a gridless algorithm for joint DOA and range estimation. Firstly, by using atomic norm denoising method and total least square estimation of signal parameters via rotational invariance techniques (TLS-ESPRIT) algorithm to get DOA and reconstruct the steering vector matrix. Next, get the paired range of targets by the least square method. Finally, the effectiveness of the algorithm is proved by experimental data of 2.4G band FMCW MIMO radar which is more excellent than 2D-MUSIC.

## 38.2 Radar Model

Assume that a FMCW MIMO radar with an equivalent virtual array, which contains $M$ equally spaced elements. The signal model of single snapshot is as follows [1]:

$$\begin{bmatrix} x_1(n) \\ x_2(n) \\ \vdots \\ x_M(n) \end{bmatrix} = \begin{bmatrix} e^{j\frac{4\pi f_c x_1 \sin\theta_1}{c}} & e^{j\frac{4\pi f_c x_1 \sin\theta_2}{c}} & \cdots & e^{j\frac{4\pi f_c x_1 \sin\theta_k}{c}} \\ e^{j\frac{4\pi f_c x_2 \sin\theta_1}{c}} & e^{j\frac{4\pi f_c x_2 \sin\theta_2}{c}} & \cdots & e^{j\frac{4\pi f_c x_2 \sin\theta_k}{c}} \\ \vdots & \vdots & \ddots & \vdots \\ e^{j\frac{4\pi f_c x_M \sin\theta_1}{c}} & e^{j\frac{4\pi f_c x_M \sin\theta_2}{c}} & \cdots & e^{j\frac{4\pi f_c x_M \sin\theta_k}{c}} \end{bmatrix} \begin{bmatrix} V_1 e^{j\left(\frac{4\pi B T_s R_1 n}{Tc} + \frac{4\pi f_c R_1}{c}\right)} \\ V_2 e^{j\left(\frac{4\pi B T_s R_2 n}{Tc} + \frac{4\pi f_c R_2}{c}\right)} \\ \vdots \\ V_k e^{j\left(\frac{4\pi B T_s R_k n}{Tc} + \frac{4\pi f_c R_k}{c}\right)} \end{bmatrix} + \begin{bmatrix} w_1(n) \\ w_2(n) \\ \vdots \\ w_M(n) \end{bmatrix} \tag{38.1}$$

where $f_c$ denotes the carrier frequency, $T$ is the cycle time, $c$ is the velocity of light, $T_s$ is the sampling time, $x_i = d \times (i-1)$ is the distance from the reference element, $d$ denote the interspacings of the virtual elements, $B$ is the bandwidth, $\theta_k$, $R_k$ are the DOA and range of the $k$th target, respectively, $V_k$ is the amplitude of the $k$th signal, and $w_i(n)$ is white noise. Combined with Eq. 38.1, the model of multisnapshot can be written as:

$$X = AS + N \in \mathbb{C}^{M\times L} \tag{38.2}$$

where the $\boldsymbol{X}$ denotes multisnapshot data matrix, the steering matrix is given by $\boldsymbol{A}$, $\boldsymbol{S}$ is the matrix of $K$ signals, $\boldsymbol{N}$ denotes the white noise matrix, and $\cdot \in \mathbb{C}^{x \times y}$ denotes $\cdot$ is a $x \times y$ matrix.

## 38.3 DOA Estimation

### *38.3.1 Atomic Norm Denoising*

According to [5], the optimization problem of atomic norm denoising method can be expressed in the form of minimization as follows:

$$\min_{\boldsymbol{Z}} \tau \|\boldsymbol{Z}\|_{\mathcal{A}} + \frac{1}{2}\|\boldsymbol{X} - \boldsymbol{Z}\|_2^2 \tag{38.3}$$

where $\|\cdot\|_{\mathcal{A}}$ denotes the atomic norm, $\boldsymbol{Z}$ represents complex matrix, and $\tau$ is the regularization coefficient. This optimization problem can be changed into the following SDP:

$$\begin{aligned} &\min_{\boldsymbol{D},\boldsymbol{Z},\boldsymbol{u}} \tfrac{\tau}{2}(\boldsymbol{D} + \boldsymbol{u}_1) + \tfrac{1}{2}\|\boldsymbol{X} - \boldsymbol{Z}\|_2^2 \\ &s.t. \begin{bmatrix} \boldsymbol{D} & \boldsymbol{Z}^{\mathrm{H}} \\ \boldsymbol{Z} & \boldsymbol{H}(\boldsymbol{v}) \end{bmatrix} \geq 0 \end{aligned} \tag{38.4}$$

where $\boldsymbol{R}_{\mathrm{o}} = \boldsymbol{H}(\boldsymbol{v})$ is a low-noise covariance matrix with the structure of Toeplitz, which is composed of vector $\boldsymbol{v} = [\boldsymbol{v}_1, \boldsymbol{v}_2, \ldots, \boldsymbol{v}_M]^{\mathrm{T}} \in \mathbb{C}^M$, and $\boldsymbol{D}$ is a positive definite symmetric matrix.

### *38.3.2 TLS-ESPRIT DOA Estimation*

The signal subspace is formed by eigen decomposition of $\boldsymbol{R}_{\mathrm{o}}$, and the $k$ largest eigenvalues are taken as the signal subspace, and the following eigenvalue decomposition is calculated after being divided into $\boldsymbol{E}_x$ and $\boldsymbol{E}_y$ two parts:

$$\boldsymbol{E}_x^{\mathrm{H}}\boldsymbol{E}_x + \boldsymbol{E}_y^{\mathrm{H}}\boldsymbol{E}_y = \boldsymbol{U}\boldsymbol{\Lambda}\boldsymbol{U}^{\mathrm{H}} \tag{38.5}$$

Decompose $\boldsymbol{U}$ into a submatrix of $k \times k$.

$$\boldsymbol{E} = \begin{bmatrix} \boldsymbol{E}_{11} & \boldsymbol{E}_{12} \\ \boldsymbol{E}_{21} & \boldsymbol{E}_{22} \end{bmatrix} \tag{38.6}$$

Calculate the eigenvalue $\lambda_i, i = 1, 2, \ldots, k$ of $\boldsymbol{\Psi} = -\boldsymbol{E}_{12}\boldsymbol{E}_{22}^{-1}$, then the estimated value of DOA is obtained by the following equation:

$$\hat{\theta}_i = \arcsin\left\{\frac{c}{4\pi f_c d}\text{angle}(\lambda_i)\right\}, \quad i = 1, 2, \ldots, k \tag{38.7}$$

## 38.4 Range Estimation

Using the DOA estimation results of the previous section, we can reconstruct the steering vector matrix of the system:

$$\breve{\boldsymbol{A}} = \left[\boldsymbol{a}(\hat{\theta}_1), \boldsymbol{a}(\hat{\theta}_2), \ldots, \boldsymbol{a}(\hat{\theta}_k)\right] \tag{38.8}$$

where $\boldsymbol{a}(\hat{\theta}_1) = \left[e^{j4\pi f_c x_1 \sin\theta_i/c}, \ldots, e^{j4\pi f_c x_M \sin\theta_i/c}\right]^T$. Multiply Eq. 38.2 to the left by $\breve{\boldsymbol{A}}^+$, owing to $\breve{\boldsymbol{A}}^+\boldsymbol{A} \approx \boldsymbol{I}_{K\times K}$, and the following equation is established:

$$\boldsymbol{Y} \approx \boldsymbol{S} + \breve{\boldsymbol{A}}^+\boldsymbol{N} \tag{38.9}$$

Let $\boldsymbol{Y}_1 = \boldsymbol{Y}^T$, which every column is a noisy signal matrix transposition. According to the signal matrix form in Eq. 38.1, it is not difficult to know that the phase difference of adjacent elements in each column is $\Delta = 4\pi BT_s/T_c$; at the same time, each element contains the phase of the first element. Therefore, the phase matrix of real value can be obtained by normalization.

$$\boldsymbol{Y}_1(:, i) = \text{Im}\{\ln(\boldsymbol{Y}_1(:, i)/\boldsymbol{Y}_1(1, i))\}, \quad i = 1, 2, \ldots, k \tag{38.10}$$

where $\text{Im}\{\cdot\}$ denotes imaginary part operator, $(:, i)$ denotes $i$th column, and $(1, i)$ denotes the first element of $i$th column.

Constructing least square matrix

$$\boldsymbol{P}_r = \begin{bmatrix} 1 & 1 & \cdots & 1 \\ \Delta & 2\Delta & \cdots & L\Delta \end{bmatrix} \tag{38.11}$$

The distance estimation corresponding to each DOA estimate can be obtained by multiplying generalized inverse of $\boldsymbol{P}_r$ to each column of normalized $\boldsymbol{Y}_1(:, 1)$.

$$\hat{R}_i = \{\boldsymbol{P}_r^+\boldsymbol{Y}_1(:, i)\}_{(2)}, \quad i = 1, 2, \ldots, k \tag{38.12}$$

where $\{\cdot\}_{(2)}$ is the second element of the column vector.

## 38.5 Simulations and Experiments

In this section, the experimental data are used to compare the performance between proposed algorithm with 2D-MUSIC [3]. The 2.4G FMCW radar parameters and two actual target parameters are shown in Tables 38.1 and 38.2. Figure 38.1 is the 2.4G FMCW MIMO radar system. Figure 38.2 shows the experiment scene. The number of snapshots is 400.

As the results shown in Fig. 38.3, the proposed algorithm is better than a traditional 2D-MUSIC.

**Table 38.1** 2.4G FMCW radar parameters

| Parameter | Value | Parameter | Value | Parameter | Value |
|---|---|---|---|---|---|
| $T_s$ | $2.5\times10^{-5}$ s | $f_c$ | 2.4 Ghz | $T$ | $2.06\times10^{-4}$ s |
| $B$ | 80 Mhz | $M$ | 16 | $d$ | 0.08 m |

**Table 38.2** Two actual target parameters

| Parameter | Value | Parameter | Value (m) |
|---|---|---|---|
| $\theta_1$ | −8.5° | $R_1$ | 41.9 |
| $\theta_2$ | 0.2° | $R_2$ | 28.1 |

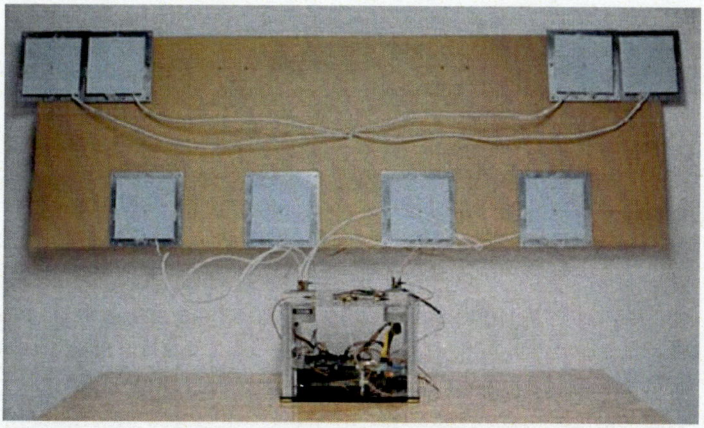

**Fig. 38.1** 2.4G FMCW MIMO radar system

**Fig. 38.2** Experiment scene

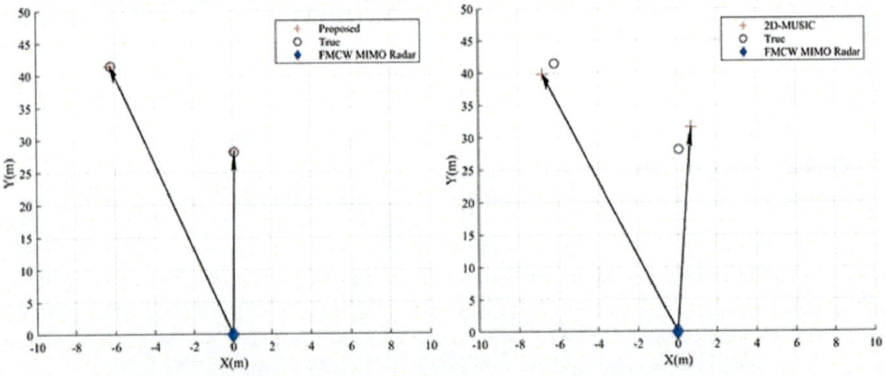


**Fig. 38.3** Comparison of target location experiments

## 38.6 Conclusion

In this work, we designed a gridless algorithm for joint DOA and range estimation, by applying atomic norm denoising method, TLS-ESPRIT algorithm, and least square method. Although the performance of the proposed method is verified on experimental data which is much better than 2D-MUSIC, it is limited to stationary narrow-band far-field targets. Therefore, in the future work, some more practical targets will be considered, such as narrow-band far-field moving targets and wide-band moving targets.

**Acknowledgements** This work is supported by Young Elite Scientists Sponsorship Program by CAST (No. 2018QNRC001), Key Research and Development Program of Hainan Province (No. ZDYF2019011), the Scientific Research Setup Fund of Hainan University (No. KYQD(ZR) 1731), the Program of Hainan Association for Science and Technology Plans to Youth R&D Innovation

(No. QCXM201706), the Scientific Research Setup Fund of Hainan University (No. KYQD(ZR) 1731), and the National Natural Science Foundation of China (No. 61701144, No. 61861015, No. 61961013).

## References

1. Huang, L., Wang, X., Huang, M., Wan, L.: An implementation scheme of range and angular measurements for FMCW MIMO radar via sparse spectrum fitting. Electronics **3**(9), 389 (2020)
2. Feger, R., Wagner, C., Schuster, S., Scheiblhofer, S., Jager, H., Stelzer, A.: A 77-GHz FMCW MIMO radar based on an SiGe singlechip transceiver. IEEE Trans. Microw. Theory Tech. **57**(5), 1020–1035 (2009)
3. Belfiori, F., Rossum Van, W., Hoogeboom, P.: 2D-MUSIC technique applied to a coherent FMCW MIMO radar. In: IET International Conference on Radar Systems (Radar 2012), pp. 1–6. IEEE, UK (2012)
4. Hamidi, S., Nezhad-Ahmadi, M., Safavi-Naeini, S.: TDM based virtual FMCW MIMO radar imaging at 79 GHz. In: 2018 18th International Symposium on Antenna Technology and Applied Electromagnetics (ANTEM), pp. 1–2. IEEE, Waterloo (2018)
5. Li, Y., Chi, Y.: Off-the-grid line spectrum denoising and estimation with multiple measurement vectors. IEEE Trans. Signal Process. **5**(64), 1257–1269 (2016)

# Author Index




L. C. Jain et al. (eds.), *Smart Communications, Intelligent Algorithms and Interactive Methods*, Smart Innovation, Systems and Technologies 257,
https://doi.org/10.1007/978-981-16-5164-9